STUDIES IN GRAPH THEORY, PART I

Studies in Mathematics

The Mathematical Association of America

Volume 1: STUDIES IN MODERN ANALYSIS
edited by R. C. Buck

Volume 2: STUDIES IN MODERN ALGEBRA
edited by A. A. Albert

Volume 3: STUDIES IN REAL AND COMPLEX ANALYSIS
edited by I. I. Hirschman, Jr.

Volume 4: STUDIES IN GLOBAL GEOMETRY AND ANALYSIS
edited by S. S. Chern

Volume 5: STUDIES IN MODERN TOPOLOGY
edited by P. J. Hilton

Volume 6: STUDIES IN NUMBER THEORY
edited by W. J. LeVeque

Volume 7: STUDIES IN APPLIED MATHEMATICS
edited by A. H. Taub

Volume 8: STUDIES IN MODEL THEORY
edited by M. D. Morley

Volume 9: STUDIES IN ALGEBRAIC LOGIC
edited by Aubert Daigneault

Volume 10: STUDIES IN OPTIMIZATION
edited by G. B. Dantzig and B. C. Eaves

Volume 11: STUDIES IN GRAPH THEORY, PART I
edited by D. R. Fulkerson

Volume 12: STUDIES IN GRAPH THEORY, PART II
edited by D. R. Fulkerson

Claude Berge
Faculté des Sciences de Paris

R. A. Brualdi
University of Wisconsin, Madison

G. B. Dantzig
Stanford University

R. J. Duffin
Carnegie-Mellon University

D. R. Fulkerson
Cornell University

R. E. Gomory
IBM T. J. Watson Research Center

Branko Grünbaum
University of Washington

A. J. Hoffman
IBM T. J. Watson Research Center

T. C. Hu
University of California, San Diego

G. J. Minty
Indiana University

C. St. J. A. Nash-Williams
University of Aberdeen

W. T. Tutte
University of Waterloo

Hassler Whitney
The Institute for Advanced Study

Studies in Mathematics

Volume 11

STUDIES IN GRAPH THEORY, PART I

D. R. Fulkerson, editor
Cornell University

Published and distributed by
The Mathematical Association of America

Library of Congress Catalog Card Number 75-24987

Complete Set ISBN 0-88385-100-8
Vol. 11 0-88385-111-3

Printed in the United States of America

Current printing (last digit):

10 9 8 7 6 5 4 3 2 1

ACKNOWLEDGMENTS

Professor Dantzig's contribution, "On the shortest route through a network", is reprinted, with slight changes, from MANAGEMENT SCIENCE, Volume 6, Number 2, January 1960; Professor Duffin's contribution, "Electrical network models", is an extension of his paper, "Network models", which appeared in SIAM-AMS PROCEEDINGS, Volume 3 (1971), pages 65–91; Professor Fulkerson's contribution, "Flow networks and combinatorial operations research", is reprinted, with slight changes, from the AMERICAN MATHEMATICAL MONTHLY, Volume 73, Number 2, February 1966; Dr. Gomory's and Professor Hu's contribution, "Multi-terminal flows in a network", is reprinted from the SIAM JOURNAL ON APPLIED MATHEMATICS, Volume 9, Number 4, December 1961; Professor Minty's contribution, "On the axiomatic foundations of the theories of directed linear graphs, electrical networks and network-programming", is reprinted, with slight changes, from the JOURNAL OF MATHEMATICS AND MECHANICS, Volume 15, Number 3 (1966); Professor Whitney's and Professor Tutte's contribution, "Kempe chains and the four colour problem", is reprinted from UTILITAS MATHEMATICA, Volume 2, November 1972.

PREFACE

It is probably fair to say, and has been said before by many others, that graph theory began with Euler's solution in 1735 of the class of problems suggested to him by the Königsberg bridge puzzle. But had it not started with Euler, it would have started with Kirchhoff in 1847, who was motivated by the study of electrical networks; had it not started with Kirchhoff, it would have started with Cayley in 1857, who was motivated by certain applications to organic chemistry, or perhaps it would have started earlier with the four-color map problem, which was posed to De Morgan by Guthrie around 1850. And had it not started with any of the individuals named above, it would almost surely have started with someone else, at some other time. For one has only to look around to see "real-world graphs" in abundance, either in nature (trees, for example) or in the works of man (transportation networks, for example). Surely someone at some time would have passed from some real-world object, situation, or problem to the abstraction we call graphs, and graph theory would have been born.

Today graph theory is a vast and somewhat sprawling subject, embracing as it does applications in many diverse areas: physics, chemistry, engineering, operations research, genetics, economics, psychology, and sociology, to name some. Dozens of books and proceedings of conferences on graph theory have appeared, mostly within the last fifteen years, and the number of journal articles dealing with graphs that have appeared in this time interval must number in the thousands. Today there are journals devoted exclusively to graph or network theory, and other journals, devoted exclusively to combinatorial mathematics, in which many, if not most, of the papers that appear are about graphs.

This recent explosion in a subject that was fairly dormant over a long period of time creates a difficult situation for one who is asked to edit a study on graph theory. Many facets of the subject must be omitted entirely; others can be treated in only a sketchy fashion. The resulting study will be biased by the editor's ignorance of some topics in the subject, and by his likes and dislikes for topics he knows something about. These remarks would apply to almost any editor; they certainly apply to me. Some of the important omissions that I know about include the fairly recent and lengthy affirmative resolution of the Heawood map conjecture by Ringel and Youngs, the solution of the Shannon switching game by Lehman, and the work of Edmonds on weighted matching theory, together with its application to very practical generalizations of the Euler problem. The latter would have brought us back to where it all started.

I shall let the papers that comprise the two volumes of this study speak for themselves. Some of them have appeared elsewhere; others appear here for the first time.

D. R. FULKERSON

CONTENTS

Pages 1–199 refer to Volume 11: Studies in Graph Theory, Part I; pages 201–413 refer to Volume 12: Studies in Graph Theory Part II.

PERFECT GRAPHS

Claude Berge

1. INTRODUCTION

Let G be a simple graph (with no loops and no multiple edges), and let $\gamma(G)$ be its *chromatic number*, i.e., the least number of colors required to color the vertices of G so that no two adjacent vertices have the same color. We shall denote by $\omega(G)$ the *clique number* of G, i.e., the maximum number of vertices in a clique. Clearly, $\gamma(G) \geqslant \omega(G)$, because if C is a maximum clique of G, any two vertices of C are joined, and all the vertices in C must have different colors. However, for many interesting classes of graphs, e.g., bipartite graphs, the equality holds.

When, for a graph, the equality holds, this does not give much information about its structure because every graph augmented with a large clique satisfies $\gamma(G) = \omega(G)$. For this reason, we need the following concept: A graph $G = (X, E)$ is said to be *γ-perfect* if, for every subset $A \subseteq X$, the induced subgraph G_A of G satisfies $\gamma(G_A) = \omega(G_A)$.

Now, denote by $\alpha(G)$ the *stability number of* G, i.e., the maximum number of vertices in a *stable set* (set of vertices that are pairwise non-adjacent). Denote by $\theta(G)$ the *partition number*, i.e., the least number of cliques whose union covers all the vertices of G.

If we consider a stable set S and a minimum partition of X into cliques, then S cannot have more than one vertex in each of these cliques; hence $\alpha(G) \leqslant \theta(G)$. For several interesting classes of graphs, e.g., bipartite graphs, the equality holds.

A graph $G = (X, E)$ is said to be *α-perfect* if for every subset $A \subseteq X$, the induced subgraph G_A satisfies $\alpha(G_A) = \theta(G_A)$.

When, in [1], we introduced these concepts, we started to investigate some classes of graphs which could be proved to be α-perfect or γ-perfect. This list immediately suggested the following conjectures:

1. (*The "weak" perfect graph conjecture*): *Is it true that a graph is α-perfect if and only if it is γ-perfect*?
2. (*The "strong" perfect graph conjecture*): *If G is a simple graph, is it true that the following three conditions are equivalent*:
   (1) *G is α-perfect*,
   (2) *G is γ-perfect*,
   (3) *G does not contain* (*as an induced subgraph*) *an odd cycle without chords, and G does not contain the complement graph of an odd cycle without chords*?

The first conjecture, which is now settled (see section 6), is interesting in itself; Fulkerson in [8], [9], gave an analytic interpretation in his theory of anti-blocking polyhedra, and reduced the conjecture to a lemma (Lemma 1 of section 6).

The second conjecture, which is stronger and still unsettled, can also be simplified. We shall show that (1) *implies* (3): If G is a graph satisfying (1), and if there exists a set A of vertices such that the induced subgraph G_A is a cycle of length $2k + 1$ without chords or, for short, $G_A \cong C_{2k+1}$, then we have

$$\alpha(G_A) = \alpha(C_{2k+1}) = k,$$

$$\theta(G_A) = \theta(C_{2k+1}) = k + 1.$$

Clearly, this contradicts (1).

Moreover, if there exists a set B of vertices such that

$G_B \cong \overline{C}_{2k+1}$, (the complement graph of C_{2k+1}), then

$$\alpha(G_B) = \omega(C_{2k+1}) = 2,$$

$$\theta(G_B) = \gamma(C_{2k+1}) = 3.$$

Again, this contradicts (1).

However, it has not been proved that (3) implies (1), or, equivalently, that (3) implies (2).

In the following sections, we shall study some classes of graphs which are both α-perfect and γ-perfect, and the reader can check easily that they fulfill condition (3).

2. COMPARABILITY GRAPHS

A graph G is said to be a *comparability graph* if it is possible to orient each edge in such a way that the relation "there is an oriented edge going from vertex a to vertex b," or in short, $a > b$, is a strict order. That is,

(1) $a > b$, $b > c$ implies $a > c$,

(2) $a > b$ implies not $b > a$.

Every induced subgraph of a comparability graph is obviously a comparability graph; a characterization of such graphs has been given independently by P. Gilmore and A. Hoffman [12] and by A. Ghouila-Houri [11].

Example.

Bipartite graph. A graph is said to be *bipartite* if it has at least one edge and if it does not contain any odd cycle. It is well known that for such a graph G, one has $\gamma(G) = 2$. If we color the vertices with two colors, and if we direct each edge from the first color to the second color, the conditions (1) and (2) are trivially true, so the graph G is a comparability graph. The converse is not true; the graph of Figure 1 is a comparability graph, but not bipartite. The reader can easily check that the graph of Figure 2 is not a comparability graph.

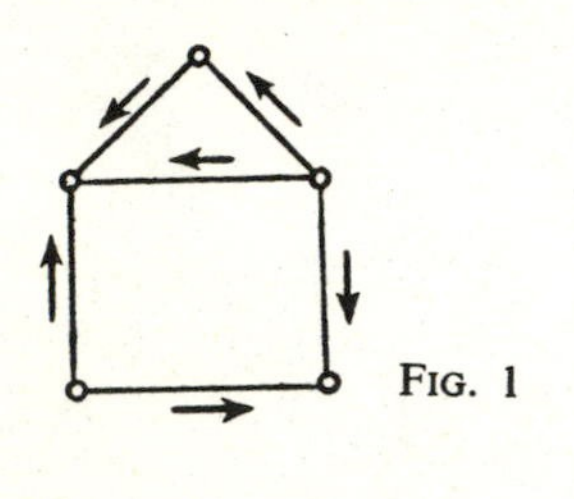

FIG. 1

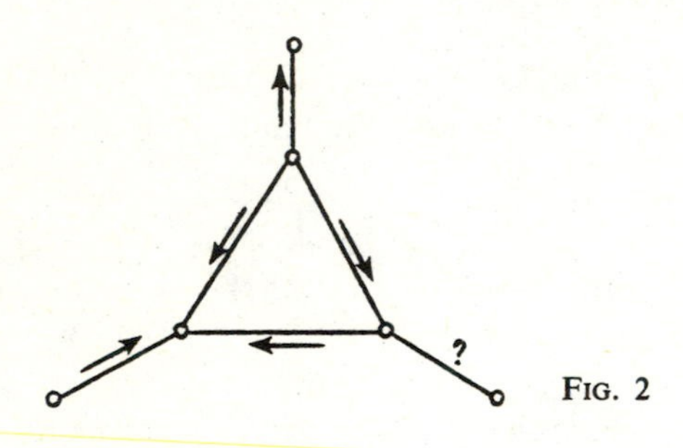

FIG. 2

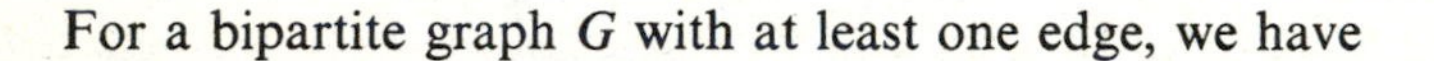

For a bipartite graph G with at least one edge, we have

$$\omega(G) = 2 = \gamma(G).$$

On the other hand, it is known that $\alpha(G) = \theta(G)$. This is equivalent to a very famous theorem of König. Thus, a bipartite graph is *perfect*.

THEOREM 1: *Every comparability graph is γ-perfect.*

Proof: Consider a graph G with an orientation $>$ which satisfies (1) and (2). Thus, the graph G has no directed circuits, and, as G is finite, to each vertex x we can assign a finite number $f(x)$ representing the length of the longest directed path issuing from x. If $\max_x f(x) = k - 1$, there exists a directed path with k vertices. There exists no clique with more than k vertices because such a clique would contain a path passing through each of its vertices (Theorem of Rédei), and the longest path contains only k vertices.

Thus, we have

$$\omega(G) = k.$$

On the other hand, consider k colors $0, 1, \ldots, k-1$, and assign color $f(x)$ to vertex x. Two adjacent vertices x and y cannot have the same color because if the edge $[x, y]$ is oriented from x to y, we have $f(x) > f(y)$; therefore,

$$\gamma(G) \leqslant k.$$

As we have always

$$\gamma(G) \geqslant \omega(G) = k,$$

we have finally

$$\gamma(G) = k = \omega(G).$$

THEOREM 2: *Every comparability graph is* α*-perfect.*

Proof: A very famous theorem of Dilworth states: "If the orientation of the edges of a graph G satisfies (1) and (2), then $\alpha(G)$ is equal to the smallest number of disjoint paths which cover all the vertices." But to each path of G corresponds a clique, and to each clique corresponds a path; therefore, we have $\alpha(G) = \theta(G)$.

3. TRIANGULATED GRAPHS

A graph is said to be *triangulated* if every cycle of length greater than three possesses a chord, that is, an edge joining two non-consecutive vertices of the cycle. Triangulated graphs arise in many contexts.

Example 1.

Interval graphs. Consider on a line a finite family of intervals, and draw a graph whose vertices represent the intervals, two vertices being joined if the two corresponding intervals intersect; such a graph is called an *interval graph*.

We shall show that *every interval graph G is triangulated.* Otherwise, there exists in G a chordless cycle $\mu = [a_1, a_2, \ldots, a_k, a_1]$ of length $k > 3$. Let $A_i = [\alpha_i, \beta_i]$ be the interval of the family that is represented by vertex a_i. Then, A_{i-1} and A_{i+1} are two disjoint intervals (since the cycle μ is chordless). If, say, A_3 is at the right-hand side of A_1, then $\beta_1 < \beta_2 < \beta_3 < \cdots < \beta_k < \beta_1$, which is a contradiction.

The problem of characterizing a graph representing a family of intervals was first put by G. Hajos [14] as follows: In a university, each student has to go once a day to the library; at the end of the day we ask each of them whom he has met, and we draw a graph G whose vertices represent the students, two vertices being joined if the two corresponding students met at the library. For each student, we have corresponding an interval of time during which he stayed at the library, and G is the representing graph of this family of intervals. Hajos later gave an algorithm to locate the intervals, and P. C. Gilmore and A. J. Hoffman [12] gave a complete characterization of interval graphs. See also Lekkerkerker and Boland [15].

Interval graphs arise also in psychology as follows: Consider on a line a finite number of points $P_1, P_2, \ldots, P_n$, and an infinite family of intervals Ω; two points, P_i and P_j, are said to be *indistinguishable* if there exists in Ω an interval which contains both P_i and P_j.

A problem which has been considered in psychosociology is to characterize the graphs of indistinguishable pairs of points. In fact, such a graph G is the representing graph of a family of intervals $I_1, I_2, \ldots, I_n$. Interval I_i corresponding to point P_i is defined as follows: I_i is the intersection of interval $[P_i, +\infty]$ with the union of all the intervals of Ω which contain P_i.

Consider two points P_i and P_j, with $P_i < P_j$. If P_i and P_j are indistinguishable, one has $P_j \in I_i$; therefore $I_i \cap I_j \neq \emptyset$. Conversely, if $I_i \cap I_j \neq \emptyset$, one has $P_j \in I_i$; therefore P_i and P_j are indistinguishable.

Thus, each graph of indistinguishable pairs of points is a representing graph of a family of intervals (and each representing graph of a family of intervals is also a graph of indistinguishable pairs of points).

Example 2.

A graph G is said to be a *cactus* if it is connected and does not possess any cycle of length greater than 3. Cacti are considered in physics and are obviously triangulated.

Example 3.

Given a graph G, its *adjoint* G^* is a graph whose vertices represent the edges of G, two vertices being joined if they represent adjacent edges of G. *The graph G^*, adjoint of a cactus G, is triangulated.* Assume that there exists in G^* a cycle $(u_1^*, u_2^*, \ldots, u_k^*)$ without a chord, with $k > 3$; it corresponds to a cycle of G with edges $u_1, u_2, \ldots, u_k$. This contradicts the fact that G is a cactus.

We must remark that "triangulated" and "comparability" graphs are two independent concepts. The graph pictured in Figure 1 is a comparability graph, but it is not a triangulated graph; on the other hand, the graph of Figure 2 is triangulated, but is not a comparability graph.

LEMMA: *If a triangulated graph G is connected and is not a clique, it contains an articulation set which is a clique.*

Proof: As G is not a clique, there exists at least two non-adjacent vertices, so there exists at least one articulation set. Let A be a minimal articulation set, whose removal creates several connected components $C, C', C'', \ldots$. Every element a of A is joined by an edge to every component because, if not, $A - \{a\}$ would be also an articulation set, and A would not be a minimal articulation set. Consider in A two distinct elements a_1 and a_2. There exists a chain

$$\mu = [a_1, c_1, c_2, \ldots, c_p, a_2], \qquad c_1, c_2, \ldots, c_p \in C,$$

and consider μ a shortest chain of that kind. There exists also a chain

$$\mu' = [a_2, c'_1, c'_2, \dots, c'_q, a_1], \qquad c'_1, c'_2, \dots, c'_q \in C',$$

and consider μ' a shortest chain of that kind. The cycle

$$\mu + \mu' = [a_1, c_1, c_2, \dots, c_p, a_2, c'_1, c'_2, \dots, c'_q, a_1]$$

does not possess any of the following chords:

$$\left.\begin{array}{r} [a_1, c_i] \text{ with } i \neq 1, \\ [c_i, c_j] \text{ with } i \neq j \pm 1, \\ [a_2, c_i] \text{ with } i \neq p, \end{array}\right\} \quad \begin{array}{l} \text{because } \mu \text{ would not} \\ \text{be a shortest chain} \end{array}$$

$$[c_i, c'_j],\} \quad \begin{array}{l} \text{because } C \text{ and } C' \\ \text{are two disjoint} \\ \text{connected components.} \end{array}$$

$$\left.\begin{array}{r} [a_2, c'_j] \text{ with } j \neq 1, \\ [c'_i, c'_j] \text{ with } i \neq j, \\ [a_i, c'_j] \text{ with } j \neq q, \end{array}\right\} \quad \begin{array}{l} \text{because } \mu' \text{ would not} \\ \text{be a shortest chain.} \end{array}$$

As the graph G is triangulated, the cycle $\mu + \mu'$ (whose length is at least 4) possesses a chord, and this chord is necessarily $[a_1, a_2]$.

Thus, every pair of vertices $a_1, a_2 \in A$ is joined, and therefore A is a clique.

THEOREM 3: (Berge [3]). *Every triangulated graph is γ-perfect.*

Proof: Assume that the theorem is true for all graphs of order less than n, and let us show that it is also true for a graph G of order n. If G is a clique, the theorem is true. If G is not a clique, it follows from the lemma that there exists an articulation set A that is a clique.

Let $C_1, C_2, \dots$ be the connected components of the subgraph

G_{X-A} obtained from G by removing A. By the induction hypothesis, we have for all i,

$$\gamma(G_{A \cup C_i}) = \omega(G_{A \cup C_i}) \leqslant \omega(G).$$

We can color with $\omega(G)$ colors the vertices of each $G_{A \cup C_i}$ separately, and this gives a coloration of G with $\omega(G)$ colors. Hence, $\gamma(G) = \omega(G)$. Q.E.D.

THEOREM 4: (Hajnal, Suranyi [13]). *Every triangulated graph is α-perfect.*

The proof is similar to the preceding one.

COROLLARY: *Every interval graph is α-perfect.*

This Corollary can be rephrased as follows: *If $(I_1, I_2, \ldots, I_n)$ is a family of intervals on the line, and if the maximum number of pairwise disjoint intervals is k, then it is possible to find k points on the line such that each interval of the family contains at least one of them.*

This result was previously obtained by T. Gallai.

4. UNIMODULAR GRAPHS

Given a graph G with vertices $x_1, x_2, \ldots, x_n$, let $C_1, C_2, \ldots, C_m$ be its maximal cliques. A matrix $M = (m_j^i)$, with n columns and m rows, is said to be the *clique-incidence* matrix of G if

$$m_j^i = \begin{cases} 0 \text{ if } x_j \notin C_i \\ 1 \text{ if } x_j \in C_i. \end{cases}$$

DEFINITION 1: A graph G is said to be unimodular if its clique-incidence matrix M is totally unimodular (that is, if every square submatrix of M has a determinant equal to 0, $+1$ or -1).

Obviously, if G is unimodular, a subgraph of G has a clique-incidence matrix which is a submatrix of M; hence this subgraph is also unimodular. An alternate definition is:

DEFINITION 2: For $A \subset X$, a clique C of a graph G is said to be even in A if $|C \cap A|$ is even; G is said to be unimodular if every non-empty subset A contains two disjoint sets A_1 and A_2 (not both empty) such that each maximal clique C, even in A, satisfies $|C \cap A_1| = |C \cap A_2|$.

The equivalence of the two definitions is a particular case of a result of A. Ghouila-Houri [11].

Example 1.

A bipartite graph is unimodular, because it is well known that the edge-incidence matrix of a bipartite graph is totally unimodular.

Example 2.

The adjoint G^* of a bipartite graph G is unimodular, because the clique-incidence matrix of G^* is nothing else than the transpose M^* of the edge-incidence matrix M of G.

Example 3.

An interval graph is unimodular. We have seen that an interval graph G can represent a family of points on a line, two points P_i and P_j being joined if and only if there exists an interval $\omega \in \Omega$ which covers both of them (see Example 1, Section 3). A set A of vertices can be ordered by the natural order of the points of the line that they represent: call A_1 the points whose order is even, and A_2 the points whose order is uneven. A maximal clique C corresponds to an interval of points on the line. Thus, if C is even in A, we have $|C \cap A_1| = |C \cap A_2|$, and, by Definition 2, the graph G is unimodular.

The concept of a unimodular graph is independent of the one of a triangulated graph, or of a comparability graph. For instance, the triangle inscribed in a hexagon (Figure 3) is triangulated but is not unimodular.

The graph of Figure 4 is unimodular, but it is not a triangulated graph, nor a comparability graph.

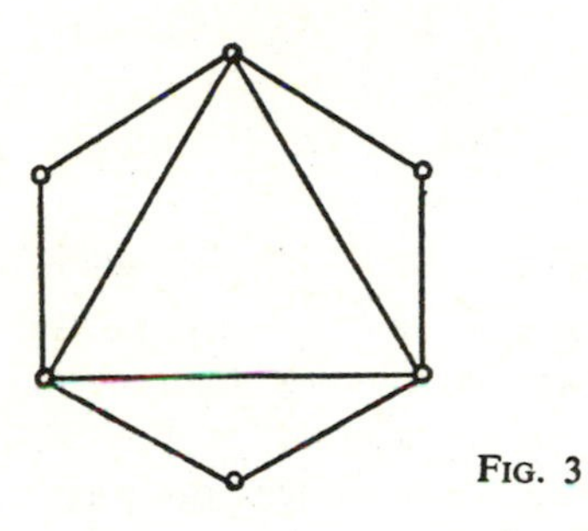

FIG. 3

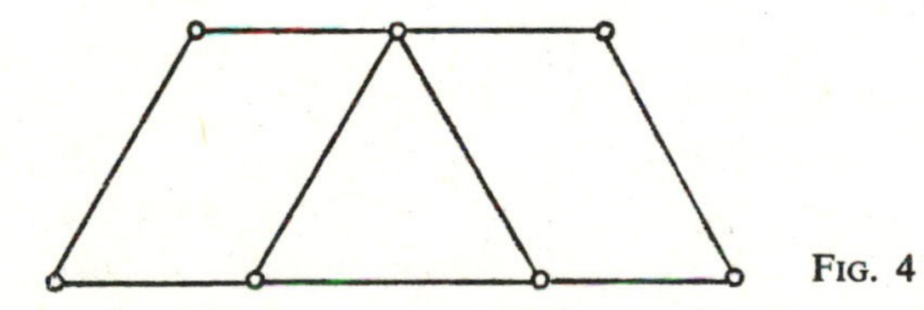

FIG. 4

THEOREM 5: *A unimodular graph G is α-perfect.*

Proof: Actually, this statement is already given in different forms by different authors. All we need is to prove that $\alpha(G) = \theta(G)$.

We shall define a stable set S by a vector

$$\alpha = (\alpha_1, \alpha_2, \ldots, \alpha_n),$$

with $\alpha_j = 1$ if $x_j \in S$, $\alpha_j = 0$ if $x_j \notin S$. A maximum stable set is given by a linear program in integers:

(1) $\alpha \geqslant 0$,

(2) $\alpha \leqslant 1 = (1, 1, \ldots, 1)$,

(3) $M\alpha \leqslant 1$,

(4) maximize $\sum_{j=1}^{n} \alpha_j$.

Condition (2) can be deleted, because it is contained in (3). The dual linear program is:

(1′) $\lambda = (\lambda_1, \lambda_2, \ldots, \lambda_m) \geqslant 0$,

(2′) $M^*\lambda \geqslant 1$,

(3′) minimize $\sum_{i=1}^{m} \lambda_i$.

We can also add

(4′) $\lambda \leqslant 1$,

because (2′) and (3′) imply (4′). In other words, we are trying to find a minimal family of cliques which cover all the vertices. As the matrix M is totally unimodular, by the duality theorem of linear programming and the fact that unimodularity implies the existence of integer solutions to the pair of dual linear programs being considered, we have:

$$\alpha(G) = \max \sum_{j=1}^{n} \alpha_j = \min \sum_{i=1}^{m} \lambda_i = \theta(G).$$

THEOREM 6: *A unimodular graph G is γ-perfect.*

Proof: All we need is to prove that $\gamma(G) = \omega(G)$. Let k be the largest number of elements in a clique. Consider the clique-incidence matrix M, and find a vector $\alpha = (\alpha_1, \alpha_2, \ldots, \alpha_n)$ which satisfies

$(0, 0, \ldots, 0) \leqslant \alpha \leqslant (1, 1, \ldots, 1)$,

$\langle M^i, \alpha \rangle = 1$, if clique C_i contains exactly k vertices, and

$\langle M^i, \alpha \rangle \leqslant 1$, if clique C_i contains less than k vertices.

The point $\alpha = (1/k, 1/k, \ldots, 1/k)$ satisfies these inequalities, hence they are consistent. But by the unimodular property of the matrix M, there is an integral solution

$$\alpha^1 = (\alpha_1^1, \alpha_2^1, \ldots, \alpha_n^1).$$

Consider the set S_1 of all vertices x_i such that $\alpha_i^1 = 1$: it is a stable set, and it meets all the cliques with k elements. Color with a first color the vertices of S_1. The subgraph obtained by deleting S_1 is also unimodular and its largest clique contains $k - 1$ elements. Color with a second color a stable set S_2 which meets all the cliques with $k - 1$ elements, etc.

By such a process, we can color all the vertices of G with k colors; hence $\gamma(G) = k = \omega(G)$.

This theorem contains a fundamental theorem about bipartite graphs, namely that the edge-chromatic number of a bipartite graph is equal to the maximum degree of its vertices.

5. AN APPLICATION IN CODING THEORY

The following problem was raised by Shannon [19].

Consider the very simple case of a transmitter which can send five signals a, b, c, d, e; at the receiving end, each signal can give rise to two different interpretations: signal a can give p or q, signal b can give q or r, etc . . . , as shown in the diagram of Figure 5. What is the maximum number of signals which can be used so that there is no possibility of confusion on reception? The problem reduces to finding a maximum stable set S of a graph G (Figure 6) where two vertices are adjacent if they represent two signals which can be confused; we take $S = \{a, c\}$ and $\alpha(G) = 2$.

In place of single letter signals, we could use "words" of two letters, on condition that these do not lead to confusion on reception. Using the letters a and c which cannot be confused, we form the code: aa, ac, ca, cc. But a richer code is: aa, bc, ce, db,

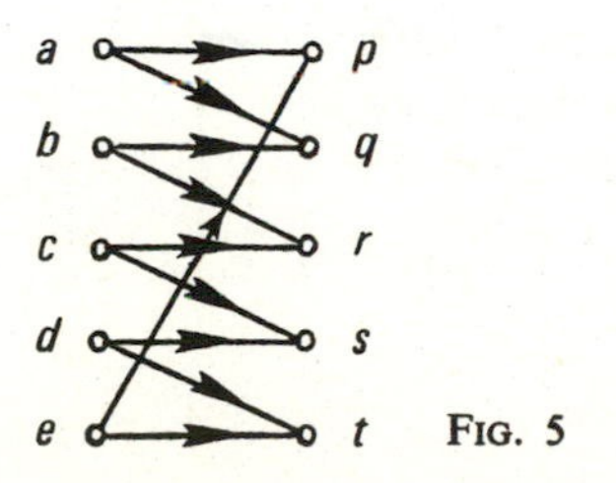

FIG. 5

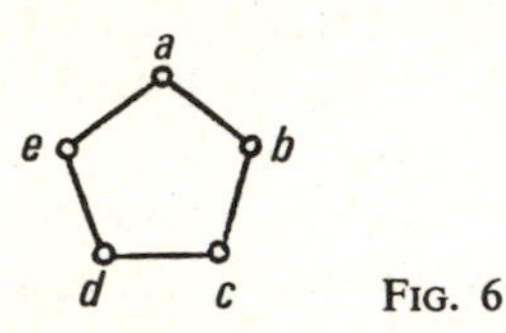

FIG. 6

ed. (It can be proved immediately that no two of these words can be confused at the receiving end.)

By definition, the *product* of two graphs $G = (X, E)$ and $H = (Y, F)$ is the graph $G \times H$ whose vertices are the pairs xy with $x \in X, y \in Y$, two vertices xy and $x'y'$ being joined if one of the following conditions holds:

(1) $x = x'$ and $[y, y'] \in F$,
(2) $[x, x'] \in E$ and $y = y'$,
(3) $[x, x'] \in E$ and $(y, y') \in F$.

With the graph G of Figure 6, two words xy and $x'y'$ can be confused if they represent two adjacent vertices of the product graph $G \times G = G^2$, and the richness of the code using two-letter words is $\alpha(G^2) = 5$; in general, the maximum number of words which cannot be confused in a code using n-letter words, is the stability number of the graph

$$G^n = G \times G \times \ldots \times G,$$

called the *n-th power of G*.

If $\alpha(G^n) > [\alpha(G)]^n$, then a code using n-letter words is better than the trivial code based upon one-letter words. Shannon raised the question of determining those graphs G which yield codes that can be improved in that sense.

THEOREM 7: *If the confusion graph G is α-perfect, then* $\alpha(G^n) = [\alpha(G)]^n$ (*and no code is better than the trivial one*).

Proof: First, we shall show that for two graphs G, H,

$$\alpha(G \times H) \geqslant \alpha(G)\alpha(H). \qquad (1)$$

Let S be a maximum stable set of G, and let T be a maximum stable set of H. Then the Cartesian product $S \times T$ is a stable set in $G \times H$. Hence

$$\alpha(G \times H) \geqslant |S \times T| = |S| \times |T| = \alpha(G)\alpha(H).$$

Now we shall show:

$$\theta(G \times H) \leqslant \theta(G)\theta(H). \tag{2}$$

Let $(C_1, C_2, \ldots, C_p)$ be a family of $p = \theta(G)$ cliques whose union covers G, and let $(D_1, D_2, \ldots, D_q)$ be a family of $q = \theta(H)$ cliques whose union covers H. Then $(C_i \times D_j / 1 \leqslant i \leqslant p, 1 \leqslant j \leqslant q)$ is a family of cliques in $G \times H$, and their union covers $G \times H$. Hence

$$\theta(G \times H) \leqslant pq = \theta(G)\theta(H).$$

From (1) and (2), we have

$$[\alpha(G)]^n \leqslant \alpha(G^n) \leqslant \theta(G^n) \leqslant [\theta(G)]^n.$$

If G is α-perfect, $\alpha(G) = \theta(G)$, and therefore

$$[\alpha(G)]^n = \alpha(G^n).$$

Q.E.D.

If, for a set of signals, the noise is "linear", that is, if the confusion graph G is an interval graph, then it follows from the Corollary to Theorem 4, that no code is better than the trivial one.

Theorem 7 also shows why the graph C_5 of Figure 6 is the only confusion graph G with less than 6 vertices, whose "capacity"

$$\sup_n \sqrt[n]{\alpha(G^n)} \quad \text{is greater than } \alpha(G).$$

The capacity of C_5 is still unknown.

6. EQUIVALENCE BETWEEN THE CONCEPTS OF α-PERFECT AND γ-PERFECT

In the preceding sections, we have proved that some classes of graphs are both α-perfect and γ-perfect. There are other similar results:

THEOREM 8: *If every odd cycle of length* $\geqslant 5$ *admits an edge such that the maximal cliques containing this edge contain at least three vertices of the cycle, then the graph is* α*-perfect* (Berge, Las Vergnas, [6]).

This Theorem generalizes Theorem 5. It can also be proved that a graph with the above property is γ-perfect (Berge, [5]).

THEOREM 9: *If every odd cycle of length* $\geqslant 5$ *possesses at least two non-crossing chords, then the graph is* α*-perfect* (Gallai, [10]).

A simpler proof of this theorem is in Suranyi, [20].

THEOREM 10: *If every odd cycle of length* $\geqslant 5$ *possesses at least two crossing chords, then the graph is* α*-perfect* (Sachs, [18]).

THEOREM 11: *If G is planar, and if no induced subgraph of G is isomorphic to* C_{2k+1} *or to* $\overline{C}_{2k+1}$, $k \geqslant 2$, *then G is* α*-perfect.* (Tucker, [21]).

Clearly, all these theorems would be trivial if the strong perfect graph conjecture were proved.

The first proof of the weak perfect graph conjecture appeared in Lovasz [16]. In [17], Lovasz proved also a statement stronger than the weak perfect graph conjecture (but weaker than the strong perfect graph conjecture). Lovasz's proofs are very closely related to earlier work of Fulkerson on anti-blocking pairs of polyhedra [8], [9]. In particular, Lemma 1 below and Fulkerson's "pluperfect" graph theorem imply the weak perfect graph conjecture, and the assumption (1) of Lemma 2 below is a special case of Fulkerson's "max-max inequality" for anti-blocking pairs of polyhedra. Before proving this result, we need two lemmas.

LEMMA 1: *Let $G = (X, E)$ be an α-perfect graph: If $x \in X$, let H be the graph obtained from G by adding a new vertex x' and by joining it to all the neighbours of x. Then H is also α-perfect.*

Proof: It suffices to show that $\alpha(H) = \theta(H)$. Let $\mathcal{C}$ be a partition of G into $\theta(G)$ cliques, and let C_x be the clique of $\mathcal{C}$ that contains x. If there exists in G a maximum stable set containing x, then

$$\alpha(H) = \alpha(G) + 1.$$

Since $\mathcal{C} \cup \{x'\}$ is a partition of H, it follows that

$$\theta(H) = \alpha(G) + 1 = \alpha(H).$$

If there does not exist in G a maximum stable set containing x, then

$$\alpha(H) = \alpha(G).$$

Since $D = C_x - \{x\}$ meets all the maximum stable sets of G,

$$\alpha(G_{X-D}) = \alpha(G) - 1.$$

Therefore,

$$\theta(G_{X-D}) = \alpha(G) - 1 = \alpha(H) - 1.$$

Thus, we can obtain a partition of H into $\alpha(H)$ cliques by taking $D \cup \{x'\}$ and the $\alpha(H) - 1$ cliques that partition $X - D$. Hence $\alpha(H) = \theta(H)$.

Q.E.D.

LEMMA 2: *Let $G = (X, E)$ be a graph, whose proper subgraphs* *are α-perfect, and such that*

$$\omega(G_A)\alpha(G_A) \geqslant |A| \qquad (A \subseteq X). \tag{1}$$

Let H be a graph obtained from G by replacing each $x_i \in X$ by a set $X_i = \{y_i^1, y_i^2, \dots\}$ and by joining y_i^s and y_j^t iff x_i and x_j are adjacent in G. Then H satisfies also (1).

Proof: Assume that H does not satisfy (1), and has the least possible number of vertices. We shall show that this leads to a contradiction.

Clearly, $\max_i |X_i| \neq 1$, and we may assume that $|X_1| = h \geqslant 2$. Let $Y = \cup X_i$. Then

$$\omega(H_{Y-X_1}) \leqslant \omega(H),$$

$$\alpha(H_{Y-X_1}) \leqslant \alpha(H).$$

Let $y_1 \in X_1$. By the minimality of H, the subgraph H_{Y-y_1} satisfies (1); hence

$$|Y| - 1 = |Y - y_1| \leqslant \omega(H_{Y-y_1})\alpha(H_{Y-y_1})$$

$$\leqslant \omega(H)\alpha(H) \leqslant |Y| - 1.$$

Therefore, the equalities hold, and we can put:

$$\omega(H_{Y-y_1}) = \omega(H) = p,$$

$$\alpha(H_{Y-y_1}) = \alpha(H) = q,$$

$$|Y| - 1 = pq.$$

H_{Y-X_1} can be obtained from G_{X-x_1} by duplicating some vertices as in Lemma 1; hence

$$\theta(H_{Y-X_1}) = \alpha(H_{Y-X_1}) \leqslant q.$$

Thus, $Y - X_1$ can be covered by q cliques of H, say $C_1, C_2, \dots, C_q$. We may assume that these cliques are pairwise disjoint and such that $|C_1| \geqslant |C_2| \geqslant \cdots \geqslant |C_q|$. Clearly,

$$|C_i| \leqslant \omega(H) = p,$$

and

$$\sum_{i=1}^{q} |C_i| = |Y| - h = pq - (h - 1).$$

Hence, $|C_i| < p$ for at most $h - 1$ values of i, and, consequently,

$$|C_1| = |C_2| = \cdots = |C_{q-h+1}| = p.$$

Let H' be the subgraph of H induced by $C_1 \cup C_2 \cup \cdots \cup C_{q-h+1} \cup \{y_1\}$. The number of vertices in H' is

$$n(H') = p(q - h + 1) + 1 < pq + 1 = |Y|.$$

By the minimality of H, H' satisfies (1); hence

$$p\alpha(H') = \omega(H')\alpha(H') \geqslant n(H') = p(q - h + 1) + 1.$$

Hence $\alpha(H') > q - h + 1$. Let S' be a stable set of H' with $q - h + 2$ vertices. Since $C_1, C_2, \ldots, C_{q-h+1}, \{y_1\}$ is a partition of H' into $q - h + 2$ cliques, $y_1 \in S'$. Therefore, $S = S' \cup X_1$ is a stable set in H, and $q = \alpha(H) \geqslant |S| = q + 2$ which is a contradiction. Q.E.D.

THEOREM 12: (Lovász, [17]). *For a graph G, the following conditions are equivalent*:

(1) $\omega(G_A)\alpha(G_A) \geqslant |A| \qquad (A \subseteq X)$

(2) $\gamma(G_A) = \omega(G_A) \qquad (A \subseteq X)$

(3) $\alpha(G_A) = \theta(G_A) \qquad (A \subseteq X)$.

Proof: (1) *implies* (2). We shall show by induction on n that all graphs of order n satisfying (1) satisfy also (2).

Let $G = (X, E)$ be a graph of order n satisfying (1). By the induction hypothesis, for $A \subseteq X$, $A \neq X$, the complement graph $\overline{G}_A$ is γ-perfect, hence G_A is α-perfect. Put $\omega(G) = p$, and let $\mathcal{S}$ be the family of all the stable sets in G.

We shall show first that there exists in G a stable set S such that $\omega(G_{X-S}) < \omega(G)$.

Otherwise, for each $S \in \mathbb{S}$, there exists a clique $C_S \subseteq X - S$ such that $|C_S| = p$. Let H be the graph obtained from the subgraph of G induced by $\cup_S C_S$ by replacing each vertex x_i by a set X_i such that $|X_i|$ = the number of C_S containing x_i. We have

$$n(H) = \Sigma|X_i| = \sum_{S \in \mathbb{S}} |C_S| = p|\mathbb{S}|,$$

$$\omega(H) \leqslant \omega(G) = p,$$

$$\alpha(H) = \max_{T \in \mathbb{S}} \left| \bigcup_{x_i \in T} X_i \right| = \max_{T \in \mathbb{S}} \sum_{S \in \mathbb{S}} |T \cap C_S| \leqslant |\mathbb{S}| - 1.$$

From these three inequalities, it follows that

$$\omega(H)\alpha(H) \leqslant p(|\mathbb{S}| - 1) < n(H).$$

This contradicts Lemma 2. Thus, there exists in G a stable set S such that

$$\omega(G_{X-S}) \leqslant \omega(G) - 1.$$

We can color G by using a first color for the vertices in S, and $\gamma(G_{X-S}) = \omega(G_{X-S})$ colors for the other vertices. Hence

$$\gamma(G) \leqslant 1 + [\omega(G) - 1] = \omega(G).$$

Hence, $\gamma(G) = \omega(G)$, and G satisfies (2).

(2) implies (1). Let G be a graph that satisfies (2). For each $A \subseteq X$, there exists a coloring $(A_1, A_2, \ldots, A_q)$ of G_A in $q = \gamma(G_A) = \omega(G_A)$ colors. Hence

$$|A| = \sum_{i=1}^{q} |A_i| \leqslant q\alpha(G_A) = \omega(G_A)\alpha(G_A).$$

Thus, (1) follows.

(1) implies (3). Let G be a graph that satisfies (1); then its

complement $\overline{G}$ satisfies (1), and, consequently, (2). Hence

$$\alpha(G_A) = \omega(\overline{G}_A) = \gamma(\overline{G}_A) = \theta(G_A).$$

Thus, (3) follows.

(3) implies (1). Let G be a graph that satisfies (3). Then $\overline{C}$ satisfies (2), and (1). Hence

$$\omega(G_A)\alpha(G_A) = \alpha(\overline{G}_A)\omega(\overline{G}_A) \geqslant |A|.$$

Q.E.D.

REFERENCES

1. Berge, C., "Färbung von Graphen, deren sämtliche bzw. deren ungerade Kreise starr sind," (Zusammenfassung), *Wiss. Z. Martin-Luther-Univ. Halle-Wittenberg Math.-Natur. Reihe*, (1961).
2. ——, *Graphs and Hypergraphs*, Amsterdam: North Holland, 1973. (French version, Paris: Dunod, 1970).
3. ——, "Les problèmes de colorations en Théorie des Graphes," *Publ. Inst. Statist. Univ. Paris*, **9** (1960), 123–160.
4. ——, "Une application de la Théorie des Graphes à un problème de codage," *Automata Theory*, London: Academic Press, 1966, 25–34.
5. ——, "Balanced matrices," *Math. Programming*, **2** (1972), 19–31.
6. ——, M. Las Vergnas, "Sur un théorème du type König pour hypergraphes," *Ann. N. Y. Acad. Sc.*, **175** (1970), 32–40.
7. Chao, Y., "On a problem of Claude Berge," *Proc. Amer. Math. Soc.*, **14** (1963), 80–82.
8. Fulkerson, D. R., "Notes on combinatorial mathematics: anti-blocking polyhedra," *R.M.* **620**/1-PR, (1970).
9. ——, "Blocking and anti-blocking pairs of polyhedra," *Math. Programming*, **1** (1971), 168–194.
10. Gallai, T., "Graphen mit triangulierbaren ungereden Vielecken," *Magyar. Tud. Akad. Mat. Fiz. Oszt. Közl.*, **7** (1962), A-3-36.

11. Ghouila-Houri, A., "Caractérisation des graphes non orientés dont on peut orienter les arêtes de manière à obtenir le graphe d'une relation d'ordre," *C.R. Acad. Sci.*, **254** (1962), 1370.

12. Gilmore, P. C., and A. J. Hoffman, "A characterization of comparability graphs and of interval graphs," *Canad. J. of Math.*, **16** (1964), 539–548.

13. Hajnal, A., and T. Suranyi, "Über die Auflösung von Graphen vollständiger Teilgraphen," *Ann. Univ. Sci. Budapest, Eötvös Sect. Math.*, **1** (1958), 113.

14. Hajos, G., "Über eine Art von Graphen," *Math. Nachr.*, **11** (1957).

15. Lekkerkerker, C. G., and J. C. Boland, "Representation of a finite graph by a set of intervals on the real line," *Fund. Math.*, **51** (1962), 45.

16. Lovász, L., "Normal hypergraphs and the perfect-graph conjecture," *Discrete Math.*, **2** (1972), 253–267.

17. ——, "A characterization of perfect graphs," *J. Combinatorial Theory*, **13** (1972), 95–98.

18. Sachs, H., *On the Berge Conjecture Concerning Perfect Graphs, Combinatorial Structures and Their Applications*, New York: Gordon and Breach, 1970, 377–384.

19. Shannon, C. E., "The zero-error capacity of a noisy channel," *IRE Trans.*, IT. **3** (1956), 3–15.

20. Suranyi, L., "The covering of graphs by cliques," *Studia Sci. Math. Hungar.*, **3** (1968), 345–349.

21. Tucker, A., "Perfect graphs and an application to refuse collection", *SIAM Rev.*, **15** (1973), 585–590.

TRANSVERSAL THEORY AND GRAPHS

Richard A. Brualdi

PREFACE

In a small community, there are a number of unmarried young ladies and gentlemen. All of the young ladies are eager to be married. If there were no other conditions, the only requirement that would have to be fulfilled in order that we satisfy these young ladies is that the number of available gentlemen be at least as great as the number of young ladies. But even eager young ladies do not enter into matrimony so hastily. Each of the young ladies would eliminate some of the gentlemen as potential spouses for reasons beknownst only to her and would, in effect, arrive at a list of gentlemen who would be regarded as suitable spouses. Now when is it possible for each of the young ladies of the community to be wed to a gentleman whom she regarded as suitable? Surely not always, for perhaps there are three young ladies whose lists each contain only two gentlemen, the same two gentlemen in each case. If not always, then under what circumstances? And when these circumstances are not present, what is the largest number of young ladies that can be accommodated?

Consider an ordinary chessboard which has 64 squares arranged in 8 rows and 8 columns, the squares being colored alternately red and black. If one has a supply of dominos (pieces which consist of 2 squares joined on a side) there are many ways to cover perfectly the squares of the chessboard with (32) dominos. By this we mean that each domino covers two squares of the board, no two dominos overlap (cover the same square) and each square is covered by some domino. Now take a pair of shears and, by cutting, remove some of the squares of the board. When is it possible to cover perfectly the squares of the "pruned board"? Obviously not always, for we need not even try if the number of squares of the "pruned board" is odd. But even if this number were even, it need not be possible, for each domino must cover one red square and one black square of the "pruned board," thus requiring that the "pruned board" have as many black squares as red squares. If we were to remove the two black squares adjacent to a red corner square, but not the red corner square itself, then no matter what we do to the rest of the board, we can never cover perfectly the resulting "pruned board" with dominos, for the red corner square can never be covered. When no perfect covering of a "pruned board" is possible, what is the largest number of dominos that can be placed on the "pruned board" with each domino covering two squares and no two dominos overlapping?

A manufacturing company makes a gadget which is broken down into n units. It employs m workers of different expertise, each of whom is required to spend a number of hours completing a certain task on each of the units. Let it be that the ith worker requires a whole number a_{ij} hours to complete his or her task on the jth unit ($1 \leqslant i \leqslant m$, $1 \leqslant j \leqslant n$). What is the fewest number of hours into which a timetable can be fitted to produce one of these gadgets so that no worker is working on two different units at the same time and no unit is being worked on by two different workers simultaneously? Now the ith worker is required to spend a total of $r_i = \Sigma_{j=1}^{n} a_{ij}$ hours to complete his work on the gadget while the jth unit requires a total of $s_j = \Sigma_{i=1}^{n} a_{ij}$ hours of attention from the workers. If p is the largest of the numbers $r_1, \ldots, r_m$, $s_1, \ldots, s_n$, then a gadget requires at least p hours to be completed. But will p hours always suffice?

At first glance, these three problems might appear to be three separate problems with no direct connections. We shall see, however, that there is a common thread running through them. Indeed, we shall show that the first two problems can be solved by appealing to a single theorem in the relatively new branch of combinatorial mathematics known as *transversal theory*. The third problem can be solved by appealing to another theorem in this subject on which the first theorem bears heavily. It is our intention in this paper to explore the subject of *transversal theory*, trying to capture its spirit, charm, and diversity without becoming too involved with technical matters. Thus we shall often forsake generality for simplicity, hopefully whetting the intellectual appetite of the reader. The connections between transversal theory and graph theory will emerge in the ensuing discussion.

1. TRANSVERSAL THEORY AND THE MARRIAGE PROBLEM

Let us take as our starting point the problem of marriage raised in the preface. Each young lady has prepared a list of gentlemen and desires to marry one of the gentlemen on her list. In mathematical terms, if G is the set of available gentlemen and L is the set of young ladies, then with each $l \in L$ there is associated a subset G_l (the list) of G. The problem of marriage then is one of choosing a $g_l \in G_l$ for each $l \in L$ in such a way that no two choices are the same (no two of the ladies claiming the same gentleman). The actual selection is called a system of distinct representatives while the set of gentlemen $\{g_l : l \in L\}$ thereby selected is called a transversal. The problem of distinct representatives can be considered relative to any family of subsets of a set and there need be no reference to marriage. The problem abstractly then is the following:

Let E be a set and let $\mathfrak{A} = (A_i : i \in I)$ be a family,* indexed by a set I, of not necessarily distinct subsets of the set E. A family $(e_i : i \in I)$ is a *system of representatives* of $\mathfrak{A}$ if $e_i \in A_i$ for each $i \in I$ and a *system of distinct representatives* if in addition the e_i are

*We shall use round brackets for families of objects (elements, sets, etc.) whereby objects may be repeated, and curly brackets for sets of objects.

distinct: $e_i \neq e_j$ for $i, j \in I$ with $i \neq j$. If $(e_i : i \in I)$ is a system of distinct representatives of $\mathfrak{A}$, then the set $\{e_i : i \in I\}$ is called a *transversal* of $\mathfrak{A}$. Thus a subset T of E is a transversal of the family $\mathfrak{A}$, provided there is a bijection* $\sigma : T \to I$ so that $e \in A_{\sigma(e)}$ for each $e \in T$. Of course, there may be several such bijections associated with the transversal T, that is, transversals corresponding to different systems of distinct representatives may be identical.

By way of illustration, let us consider the family $\mathfrak{A} = (A_1, A_2, A_3, A_4)$ where

$$A_1 = \{1, 2\}, \quad A_2 = \{1, 3, 4\}, \quad A_3 = \{3, 4, 5\}, \quad A_4 = \{2, 5\}.$$

Then the set $\{1, 2, 3, 5\}$ is a transversal of this family $\mathfrak{A}$, since

$$1 \in A_1, \quad 3 \in A_2, \quad 5 \in A_3, \quad 2 \in A_4.$$

The family $(1, 3, 5, 2)$ is a system of distinct representatives associated with this transversal. The family $(2, 1, 3, 5)$ is a system of distinct representatives, since

$$2 \in A_1, \quad 1 \in A_2, \quad 3 \in A_3, \quad 5 \in A_4,$$

which also gives rise to the transversal $\{1, 2, 3, 5\}$.

A family of non-empty sets always has a system of representatives since no attention need be paid to repetition. The family $\mathfrak{B} = (B_1, B_2, B_3, B_4)$ of non-empty sets where

$$B_1 = \{1, 2\}, \quad B_2 = \{2, 3, 4, 5\}, \quad B_3 = \{2\}, \quad B_4 = \{1, 2\},$$

however, has no system of distinct representatives or transversal, since the sets B_1, B_3, B_4 have among them only two elements while three would be necessary for a system of distinct representatives. Indeed, we now can formulate one prerequisite for a family $\mathfrak{A} = (A_i : i \in I)$ of sets to have a transversal and that is that any k members of the family must contain between them at least k distinct elements. If we let $|X|$** represent the cardinal number of a

*A map which is both injective (one-to-one) and surjective (onto).

**Usually $|X|$ can be taken to be one of $0, 1, 2, \ldots, \infty$ where $n < \infty$ for $n = 0, 1, \ldots$.

set X, then we can formalize this as follows: if the family $(A_i : i \in I)$ is to have a transversal, then

$$\left| \bigcup_{i \in J} A_i \right| \geqslant |J|$$

for each $J \subseteq I$. What is surprising is that this condition* is also sufficient for the existence of a transversal, and we are going to prove this momentarily. Before doing so, however, we need to bring out into the open an assumption which is implicit in our formulation of the marriage problem. This is that the number of members in the families considered is finite. We shall call a family $(A_i : i \in I)$ of sets a *finite family* if the indexing set I is a finite set. This is to be so irrespective of whether the individual sets A_i are themselves finite or infinite. If we wish, we can take the indexing set of a finite family to be $\{1, 2, \ldots, n\}$ for some positive integer n. We shall have more to say about infinite families, families where the indexing set is infinite, later.

THEOREM 1.1. *The finite family* $\mathfrak{A} = (A_i : i \in I)$ *of subsets of a set E has a transversal if and only if*

$$\left| \bigcup_{i \in J} A_i \right| \geqslant |J| \qquad (J \subseteq I). \tag{1.1}$$

This theorem was proved by P. Hall [14] in 1935 but the roots of the theorem predate Hall. Indeed, the theorem can be deduced from a result about bipartite graphs which was proved by D. König [19] several years earlier. We shall discuss König's theorem in the next section, so let us now turn to proving Theorem 1.1, which is commonly called *the marriage theorem*. The method of proof is due to Halmos and Vaughn [16] and has been used quite extensively in transversal theory. We have already remarked on

*If $|I| = n < \infty$, there are $2^n - 1$ relations to be satisfied (the condition is trivial for $J = \emptyset$) and these conditions are independent. To see this, let J_0 be a non-empty subset of $I = \{1, 2, \ldots, n\}$ and define $A_i = \{1, 2, \ldots, |J_0| - 1\}$ for $i \in J_0$ and $A_i = \{1, 2, \ldots, n\}$ otherwise.

the necessity of condition (1.1) for the existence of a transversal, so we now concentrate on proving its sufficiency and this is done by induction on $n = |I|$. We take $I = \{1, 2, \ldots, n\}$. If $n = 1$, (1.1) says that A_1 has at least one element e_1 and $\{e_1\}$ is a transversal. Now suppose $n > 1$ and assume the theorem holds for every family with fewer than n members. We distinguish two cases.

For the first case, suppose that

$$\left|\bigcup_{i\in J} A_i\right| \geqslant |J| + 1$$

for all $J \subseteq \{1, \ldots, n\}$ with $1 \leqslant |J| \leqslant n - 1$. Then choose any $e_1 \in A_1$ and consider the family $(A_i' : 2 \leqslant i \leqslant n)$ where $A_i' = A_i \setminus \{e_1\}$.* If $\emptyset \neq J \subseteq \{2, \ldots, n\}$, then since $1 \leqslant |J| \leqslant n - 1$,

$$\begin{aligned}\left|\bigcup_{i\in J} A_i'\right| &= \left|\left(\bigcup_{i\in J} A_i\right)\setminus\{e_1\}\right| \\ &\geqslant \left|\bigcup_{i\in J} A_i\right| - 1 \\ &\geqslant |J| + 1 - 1 = |J|.\end{aligned}$$

From the inductive assumption, we conclude there are distinct elements $e_2, \ldots, e_n$ with $e_2 \in A_2', \ldots, e_n \in A_n'$. Thus $\{e_1, e_2, \ldots, e_n\}$ is a transversal of $\mathfrak{A}$.

The second case is then that there exists a $J_0 \subseteq \{1, \ldots, n\}$ with $1 \leqslant |J_0| \leqslant n - 1$ such that

$$\left|\bigcup_{i\in J_0} A_i\right| = |J_0|.$$

The notation will be simpler if we take $J_0 = \{1, \ldots, k\}$ where

*If X and Y are two sets, $X \setminus Y = \{x : x \in X \text{ and } x \notin Y\}$.

$1 \leqslant k \leqslant n-1$. Let then $F = \cup_{i=1}^{k} A_i$ so that $|F| = k$ and consider the two families $(A_i : 1 \leqslant i \leqslant k)$ and $(A_i \backslash F : k+1 \leqslant i \leqslant n)$. The first of these two families, having as members some of the members of $\mathfrak{A}$, satisfies $|\cup_{i \in J} A_i| \geqslant |J|$ for all $J \subseteq \{1, \ldots, k\}$ and thus by the inductive hypothesis has a transversal which must be F because of cardinality considerations. If the second family also has a transversal, say T, then surely $T \cap F = \emptyset$ and then $F \cup T$ is a transversal of $\mathfrak{A}$. So consider any $K \subseteq \{k+1, \ldots, n\}$ and calculate that

$$\left| \bigcup_{i \in K} (A_i \backslash F) \right| = \left| \left(\bigcup_{i \in K \cup J_0} A_i \right) \backslash F \right|$$

$$= \left| \bigcup_{i \in K \cup J_0} A_i \right| - |F|$$

$$\geqslant |K \cup J_0| - k$$

$$= |K| + k - k = |K|.$$

Invoking the inductive assumption again, we conclude that the family $(A_i \backslash F : k+1 \leqslant i \leqslant n)$ has a transversal, and because of our previous remarks we have proved the theorem.

We have seen that a family of sets need not have a system of distinct representatives or transversal. Under such circumstances, it is natural to ask how many sets of the family can be represented by distinct elements. In the marriage problem formulation we are asking for the largest number of young ladies that can be married, each marrying a gentleman on her list. If $\mathfrak{A} = (A_i : i \in I)$ is a family of subsets of a set E, then a subset P of E is a *partial transversal* of $\mathfrak{A}$ provided there is an injection $\sigma : P \to I$ with $e \in A_{\sigma(e)}$ for each $e \in P$. The set P is then a transversal of the subfamily $\mathfrak{A}' = (A_i : i \in I')$ of $\mathfrak{A}$ where $I' = \{\sigma(e) : e \in P\}$.

Theorem 1.1 can be strengthened to give necessary and sufficient conditions for a family to have a partial transversal of a prescribed finite cardinal number.

THEOREM 1.2. *Let $\mathfrak{A} = (A_i : i \in I)$ be a finite family of subsets of a set E and let r be a positive integer with $r \leqslant |I|$. Then $\mathfrak{A}$ has a partial transversal of cardinality r if and only if*

$$\left|\bigcup_{i\in J} A_i\right| \geqslant |J| - (|I| - r) \qquad (J \subseteq I). \tag{1.2}$$

Observe that (1.2) is automatically satisfied if $|J| \leqslant |I| - r$. Note also that when $r = |I|$, the theorem reduces to Theorem 1.1.

To prove this theorem, let F be a set of cardinality equal to $|I| - r$ which is disjoint from E. Consider the family $(A_i^* : i \in I)$ where $A_i^* = A_i \cup F (i \in F)$. Then an easy mental exercise establishes that $(A_i : \iota \in I)$ has a partial transversal of cardinality r if and only if $(A_i^* : i \in I)$ has a transversal. The latter holds, according to Theorem 1.1, if and only if

$$\left|\bigcup_{i\in J} A_i^*\right| \geqslant |J| \qquad (J \subseteq I). \tag{1.3}$$

But for $J \neq \emptyset$, $\cup_{i\in J} A_i^* = (\cup_{i \in J} A_i) \cup F$. Thus (1.3) is equivalent to

$$\left|\bigcup_{i\in J} A_i\right| + |F| \geqslant |J| \qquad (J \subseteq I)$$

or

$$\left|\bigcup_{i\in J} A_i\right| + (|I| - r) \geqslant |J| \qquad (J \subseteq I),$$

which is equivalent to (1.2).

COROLLARY 1.3. *If $\mathfrak{A} = (A_i : i \in I)$ is a finite family of subsets of a set E, then the maximum cardinality of a partial transversal of $\mathfrak{A}$ equals*

$$\min\left\{\left|\bigcup_{i\in J} A_i\right| + |I \setminus J| : J \subseteq I\right\}.$$

This corollary, which follows immediately from the theorem, solves the problem of marriage posed in the preface.

As a further consequence of the theorem there is the following:

COROLLARY 1.4. *Let $\mathfrak{A} = (A_i : i \in I)$ be a finite family of subsets of a set E and let $P \subseteq E$. Then P is a partial transversal of $(A_i : i \in I)$ if and only if*

$$\left|\left(\bigcup_{i \in J} A_i\right) \cap P\right| + |I \backslash J| \geqslant |P| \qquad (J \subseteq I).$$

If P is a partial transversal of $\mathfrak{A}$, then $|P| \leqslant |I|$; on the other hand, if the above condition is satisfied, then by taking $J = \emptyset$, we see that $|P| \leqslant |I|$. The corollary now follows from Theorem 1.2 using the observation that P is a partial transversal of $\mathfrak{A}$ if and only if the family $(A_i \cap P : i \in I)$ has a partial transversal of cardinality equal to $|P|$.

Using the same method, we could also write down criteria for the set P to contain a partial transversal of some prescribed cardinality r.

There are a number of other extensions of Theorem 1.1 that can be obtained by suitably modifying a given family of sets. We mention only one more. The reader who is interested in pursuing this can consult Ford and Fulkerson [11] where many of these results are derived from their important max flow-min cut network flow theorem by modifying a network, or Mirsky [23].

THEOREM 1.5. *Let $\mathfrak{A} = (A_i : i \in I)$ be a finite family of subsets of a set E. Let there be given a family $(n_i : i \in I)$ of non-negative integers. Then there exists a family $(B_i : i \in I)$ of pairwise disjoint sets with $B_i \subseteq A_i$ and $|B_i| = n_i$ $(i \in I)$ if and only if*

$$\left|\bigcup_{i \in J} A_i\right| \geqslant \sum_{i \in J} n_i \qquad (J \subseteq I).$$

To obtain this theorem from Theorem 1.1, define a new family which is obtained from the family $(A_i : i \in I)$ by 'repeating' A_i n_i

times $(i \in I)$. Formally let $K_i (i \in I)$ be a collection of pairwise disjoint sets with $|K_i| = n_i (i \in I)$ and let $K = \cup_{i \in I} K_i$. Define a family $\mathfrak{A}^* = (A_k^* : k \in K)$ by $A_k^* = A_i$ if $k \in K_i$. Then the desired family $(B_i : i \in I)$ exists if and only if the family $\mathfrak{A}^*$ has a transversal. By Theorem 1.1, the latter is the case if and only if

$$\left| \bigcup_{k \in L} A_k^* \right| \geqslant |L| \qquad (L \subseteq K). \tag{1.4}$$

But if $J = \{i \in I : L \cap K_i \neq \emptyset\}$, then $\cup_{k \in L} A_k^* = \cup_{i \in J} A_i$. Thus the left side of the inequality (1.4) depends only on J. But the largest subset of K which gives rise to the same J is $\cup_{i \in J} K_i$, and this set has cardinality equal to $\Sigma_{i \in J} n_i$. Thus (1.4) is equivalent to

$$\left| \bigcup_{i \in J} A_i \right| \geqslant \sum_{i \in J} n_i \qquad (J \subseteq I),$$

and this proves the theorem.

If we return to the setting of the marriage problem where the sets A_i represent the sets of desirable gentlemen on the lists of the various young ladies, then in a polyandrous society Theorem 1.4 gives criteria for each young lady to have the number of husbands she wishes (all taken from her list) with the restriction that no gentleman have more than one wife.

Theorem 1.2 can be generalized to infinite families of sets provided we still seek only a partial transversal of some finite cardinal number r. The theorem would then read that the family $(A_i : i \in I)$ has a partial transversal of cardinality r if and only if

$$\left| \bigcup_{i \in J} A_i \right| + |I \backslash J| \geqslant r$$

for each subset J of I with $I \backslash J$ finite. The problem of finding criteria for an infinite family of sets to have a transversal will be considered in section 4.

2. BIPARTITE GRAPHS AND THE CHESSBOARD PROBLEM

The ideas and theorems in section 1 have equivalent formulations in the theory of bipartite graphs which we want to develop in this section. We also investigate the problem in the preface concerned with the placing of dominos on a "pruned chessboard" with no two of the dominos overlapping. New problems are suggested which we also consider. We first make a few remarks about graphs in general.

Let G be a *graph*. Thus G has a non-empty set N of *nodes* along with a set Δ of *edges* each of which is a set of two nodes. The graph G is a *finite graph* if its set of nodes is a finite set. We speak about two nodes x, y being *adjacent* if $e = \{x, y\}$ is an edge. Nodes x, y are then said to be *incident* with the edge e and the edge e is said to *join* x and y. The *nodes* x, y are *the nodes of the edge* $e = \{x, y\}$. A *path* P in the graph G is a sequence $x_1, x_2, \ldots, x_n$ of $n \geqslant 2$ distinct nodes such that $e_1 = \{x_1, x_2\}, \ldots, e_{n-1} = \{x_{n-1}, x_n\}$ are all edges of the graph. The path P is said to *join* x_1 and x_n; $x_1, \ldots, x_n$ are the nodes of the path while we refer to $e_1, \ldots, e_{n-1}$ as the edges of the path. A *cycle* is defined like a path except $x_n = x_1$. The graph G is *connected* if either it has exactly one node or it has at least two nodes and for all $x, y \in N$ with $x \neq y$ there is a path joining x and y.

If G is a graph with node set N and edge set Δ and $A \subseteq N$, then G_A is the graph with node set A whereby two nodes x, y of A are joined by an edge in G_A if and only if they are in G. G_A is called the *subgraph* of G *induced* by A. If a graph G is not connected, then the set N of nodes can be uniquely partitioned into sets $N_i (i \in I)$ such that G_{N_i} is connected $(i \in I)$ and no two nodes in different N_i's are adjacent. The graphs G_{N_i} are called the (connected) *components* of G.

A set M of edges of a graph G which are pairwise disjoint is called a *matching* in G. If A is a set of nodes such that each node of A is incident with an edge in M, then A is said to *meet the matching* M and M is said to *meet the set of nodes* A. Thus if A meets the matching M, then so does any subset of A. At the present time, we are only interested in matchings in bipartite graphs.

A *bipartite graph* is a graph in which the set of nodes can be partitioned into two subsets X and Y with every edge joining a node in X and a node in Y. Thus every edge e is of the form $e = \{x, y\}$ with $x \in X$ and $y \in Y$. This partitioning will be unique if and only if the bipartite graph is connected. If Δ is the set of edges of the bipartite graph, then we designate the graph as $\langle X, \Delta, Y \rangle$ or $\langle Y, \Delta, X \rangle$. Let $M = \{e_i : i \in I\}$ be a matching in $\langle X, \Delta, Y \rangle$ where $e_i = \{x_i, y_i\}$ with $x_i \in X$, $y_i \in Y (i \in I)$, and let $X' = \{x_i : i \in I\}$, $Y' = \{y_i : i \in I\}$. Then we say that M *matches* X' and Y' and that X' and Y' are *matched*.

Now let $\mathfrak{A} = (A_i : i \in I)$ be a family of subsets of a set E. Then we may associate with this family the bipartite graph $\langle I, \Delta, E \rangle$ where for $i \in I$, $e \in E$, $\{i, e\}$ is an edge if and only if $e \in A_i$. This bipartite graph contains all the pertinent information about the family $\mathfrak{A}$, for from it one can determine the elements which belong to each set of the family. Conversely, given a bipartite graph $\langle X, \Delta, Y \rangle$, we can associate two families of sets. One is the family $\mathfrak{A} = (A_x : x \in X)$ where for each $x \in X$, $A_x = \{y : \{x, y\} \in \Delta\}$; the other is the family $\mathfrak{B} = (B_y : y \in Y)$ where for each $y \in Y$, $B_y = \{x : \{x, y\} \in \Delta\}$. Two families $\mathfrak{A}$ and $\mathfrak{B}$ which arise in this way from one bipartite graph are called *dual families* (and each is called the *dual* of the other). Thus given a family $(A_i : i \in I)$ of subsets of a set E we can construct the bipartite graph (I, Δ, E) as above and then the dual family $(A_e^* : e \in E)$ where $A_e^* = \{i \in I : e \in A_i\}$.

Suppose now M is a matching in the bipartite graph $\langle I, \Delta, E \rangle$ associated with the family $\mathfrak{A} = (A_i : i \in I)$ of subsets of E. If $M = \{\{i_1, e_1\}, \ldots, \{i_p, e_p\}\}$, then $i_1, \ldots, i_p$ are distinct elements of I, $e_1, \ldots, e_p$ are distinct elements of E and $e_k \in A_{i_k} (1 \leqslant k \leqslant p)$. It follows that $(e_k : 1 \leqslant k \leqslant p)$ is a system of distinct representatives of the subfamily $(A_{i_k} : 1 \leqslant k \leqslant p)$ so that $\{e_1, \ldots, e_p\}$ is a partial transversal of $\mathfrak{A}$. Conversely, a system of distinct representatives of a subfamily of $\mathfrak{A}$ gives rise to a matching in the bipartite graph $\langle I, \Delta, E\rangle$. We can then conclude the following. The partial transversals of the family $\mathfrak{A}$ are precisely those subsets of E which are matched in $\langle I, \Delta, E\rangle$ with some subset of I; those subsets J of I which are matched in $\langle I, \Delta, E\rangle$ with some subset of E are characterized by the provision that the subfamily $(A_i : i \in J)$ have a transversal. Thus the family $\mathfrak{A} = (A_i : i \in I)$ has a transversal if and only if I is matched in $\langle I, \Delta, E\rangle$ with some subset of E. If $\mathfrak{A}$ is a finite family with $|I| = n$, then $\mathfrak{A}$ has a transversal if and only if there is a matching in (I, Δ, E) of cardinality n. This discussion serves to point out that the study of systems of distinct representatives and partial transversals of families of sets is equivalent to the study of matchings and sets that are matched in bipartite graphs.

The theorem we are about to prove was established by König [19] prior to the establishment of Theorem 1.1 by Hall. Nevertheless, Hall's theorem can be derived easily from König's theorem. We shall use Corollary 1.3 to derive König's theorem, although it is possible to give an *ad hoc* proof with induction as the principal tool. Before doing so, we need one additional concept. A set S of nodes of a bipartite graph $\langle X, \Delta, Y\rangle$ is said to be a *separating set* (or separates X and Y) provided every edge in Δ has at least one of its nodes in S.

THEOREM 2.1. *Let $\langle X, \Delta, Y\rangle$ be a finite bipartite graph. Then the maximum cardinality α of a matching equals the minimum cardinality β of a separating set.*

To prove this, let M be a matching and S any separating set of nodes. Each edge $e \in M$ has one of its nodes $s_e \in S$. Since no two edges in M have a node in common, $s_e \neq s_f$ for $e, f \in M$ with $e \neq f$. Hence $|M| \leqslant |S|$. Because this is true for every matching M and every separating set S, $\alpha \leqslant \beta$. It suffices now to show there is a separating set with cardinality α. But α equals the maximum cardinality of a partial transversal of the family $(A_x : x \in X)$ of subsets of Y associated with the given bipartite graph. Moreover by Corollary 1.3 there exists $X_1 \subseteq X$ such that

$$\alpha = \left| \bigcup_{x \in X_1} A_x \right| + |X \setminus X_1|.$$

But then $(\cup_{x \in X_1} A_x) \cup (X \setminus X_1)$ is a separating set, for if $\{x, y\}$ is an edge with $x \in X, y \in Y$ then either $x \in X \setminus X_1$ or $x \in X_1$, so that $y \in \cup_{x \in X_1} A_x$. Since this separating set has cardinality α, the theorem is proved.

Let us now return to the problem of dominos on a "pruned chessboard." We take any $m \times n$ board whose mn squares are arranged in m rows and n columns. We assume the squares are colored alternately red and black. Thus if two squares have a side in common, they are colored differently. We now cut out some of the squares of the board leaving a "pruned board." We let R denote the set of squares of the "pruned board" which are colored red and B those colored black. We associate with this "pruned board" a bipartite graph $\langle R, \Delta, B \rangle$ where for $r \in R$, $b \in B$, $\{r, b\}$ is an edge if and only if r and b are adjacent squares (that is, have a side in common). Thus r and b are joined by an edge if and only if a domino can cover both squares simultaneously. Of course every domino is to cover two squares of the "pruned board" and these must have different colors. The two families of sets associated with this bipartite graph are the families $(B_r : r \in R)$ where for $r \in R$, B_r is the set of all black squares of the "pruned board" which are adjacent to r and the family $(R_b : b \in B)$ where for $b \in B$, R_b is the set of all red squares of the "pruned board" which are adjacent to b.

Now suppose we have a collection $D_1, \ldots, D_t$ of non-overlapping dominos on the "pruned board." If D_i covers red square r_i and black square b_i, then $r_1, \ldots, r_t$ are distinct and so are $b_1, \ldots, b_t$. Moreover, $\{r_i, b_i\}$ is an edge of the bipartite graph $\langle R, \Delta, B \rangle$ $(1 \leqslant i \leqslant t)$. Hence the set of edges $\{\{r_1, b_1\}, \ldots, \{r_t, b_t\}\}$ is a matching of cardinality t in the bipartite graph. Conversely, any matching of cardinality t corresponds to a set of t non-overlapping dominos on the 'pruned board.' We can now apply our previous results to this situation to arrive at the following conclusions:

(a) *It is possible to cover perfectly the squares of a "pruned board" with dominos* if and only if the number of red squares equals the number of black squares and in the associated bipartite graph* $\langle R, \Delta, B\rangle$ *no set of fewer than* $|R|(=|B|)$ *nodes is a separating set. Alternately, if for any set* R_0 *of red squares,* $B(R_0)$ *is the set of black squares which are adjacent to at least one red square in* R_0*, the latter condition on separating sets can be replaced by*

$$|B(R_0)| \geqslant |R_0| \qquad (R_0 \subseteq R).$$

(b) *The maximum number of non-overlapping dominos that can be placed on the "pruned board" equals the minimum cardinality of a separating set in the associated bipartite graph* $\langle R, \Delta, B\rangle$*. Alternately it equals*

$$\min\{|B(R_0)| + |R \setminus R_0| : R_0 \subseteq R\}.$$

The two formulations in (a) and in (b) arise from the two points of view we have elucidated, bipartite graphs or families of sets.

Let us now consider another question relative to a "pruned board." A set of dominos on a "pruned board" is said to *cover* the squares of the board provided every square is covered by at least one domino. In contrast to a perfect cover, the dominos are allowed to overlap. What is the smallest number of dominos that will cover the squares of a "pruned board?" If as before R denotes the red squares of the "pruned board" and B the black squares, then at least $\max\{|R|, |B|\}$ dominos are needed to cover. But this number won't always suffice, as the pruned board in Figure 1 shows.

Of course we need to assume that each square is adjacent to some square of the "pruned board," for otherwise no set of dominos can ever cover. To answer the above question, we shall

---

*Recall that this means that each domino covers two squares of the 'pruned board' and every square is covered by exactly one domino.

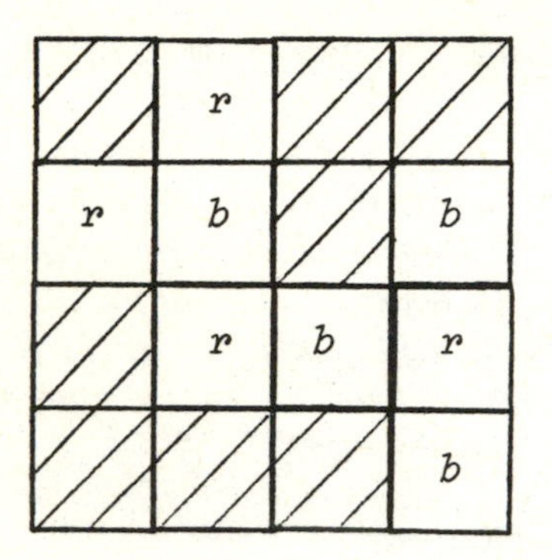

FIG. 1

invoke another theorem of König [19] about bipartite graphs. Rather than prove this theorem directly, we shall derive a theorem of Gallai [13] about graphs in general and then deduce the second theorem of König from the first and Gallai's theorem.

A set A of nodes of a graph is called (internally) *stable* (also called independent) provided no edge of the graph has both its nodes in A. Let α_0 denote the maximum cardinality of a stable set of nodes. It should be realized that this is not necessarily the same as the cardinality of a maximal stable set of nodes. For example the graph with three nodes x, y, z and edges $\{x, y\}$, $\{x, z\}$ has $\alpha_0 = 2$; yet $\{x\}$ is a maximal stable set in the sense that no set of nodes which properly contains $\{x\}$ is stable. A set C of nodes of a graph is said to *cover the edges** of the graph if every edge has at least one of its nodes in C. We let β_0 equal the minimum cardinality of a set of nodes that cover the edges of the graph. Similarly, a set F of edges of the graph is said to *cover the nodes* of the graph provided every node is incident with at least one edge in F. We set β_1 equal to the minimum cardinality of a set of edges that cover the nodes. Lastly, we set α_1 equal to the maximum cardinality of a matching in G. We can now state and prove Gallai's theorem.

---

*This coincides with the notion of a separating set of nodes that we defined for a bipartite graph. It is more common to use the phrase 'cover the edges' in the context of general graphs.

THEOREM 2.2. *Let G be a finite graph with n nodes in which each node is incident with at least one edge. Then*

$$\alpha_0 + \beta_0 = n = \alpha_1 + \beta_1.$$

Let N be the set of nodes of G so that $n = |N|$. If A is a stable set of nodes with $|A| = \alpha_0$, then every edge of G has at least one of its nodes in $N \setminus A$. Thus $N \setminus A$ covers the edges of G so that

$$n - \alpha_0 = |N \setminus A| \geqslant \beta_0$$

or

$$n \geqslant \alpha_0 + \beta_0.$$

To obtain the reverse inequality, let B be a set of nodes which cover the edges of G with $|B| = \beta_0$. Then since every edge has at least one of its nodes in B, no two nodes of $N \setminus B$ are joined by an edge. Thus $N \setminus B$ is a stable set of nodes so that

$$n - \beta_0 = |N \setminus B| \leqslant \alpha_0$$

or

$$n \leqslant \alpha_0 + \beta_0.$$

We conclude that $\alpha_0 + \beta_0 = n$.

To obtain the other identity, let M be a matching with $|M| = \alpha_1$. For each node which is not a node of one of the edges in M choose any edge incident with it. No two choices can be the same, for if this happened there would be an edge e having no node in common with any edge in M making $M \cup \{e\}$ a matching of cardinality $\alpha_1 + 1$. These edges along with the edges in M thus comprise a set of edges which cover the nodes of G with cardinality equal to $\alpha_1 + (n - 2\alpha_1)$ or $n - \alpha_1$. Thus $\beta_1 \leqslant n - \alpha_1$ or $\alpha_1 + \beta_1 \leqslant n$. To obtain the reverse inequality, take a set D of edges which cover the nodes of G with $|D| = \beta_1$. If $e = \{x, y\}$ is any edge in D, then not both x and y can be incident with other edges in D, for if they were, $D \setminus \{e\}$ would still cover all nodes and $|D \setminus \{e\}| = \beta_1 - 1$. Let $a_1, \ldots, a_p$ be those nodes of G which are incident with two or more edges of D (of course there need not be any such nodes in which case D is a matching, and $\alpha_1 = \beta_1 = n/2$). Let the set of edges in D which are incident with a_i be $D(a_i)$. If $\{a_i, b\}$ is an edge in $D(a_i)$, then since a_i is incident with at least two edges of D, the edge $\{a_i, b\}$ is the only edge of D incident with b. For $i = 1, \ldots, p$ choose an edge in $D(a_i)$. Then these p edges along with all edges in $D \backslash (\cup_{i=1}^{p} D(a_i))$ form a matching M in G and thus $|M| \leqslant \alpha_1$. If A is the set of all nodes incident with edges in $\cup_{i=1}^{p} D(a_i)$, then

$$|D| = |A| - p + \frac{n - |A|}{2}, \quad |M| = \frac{n - |A|}{2} + p$$

so that

$$|D| + |M| = n.$$

Since $|M| \leqslant \alpha_1$ and $|D| = \beta_1$, we conclude that $\alpha_1 + \beta_1 \geqslant n$. Thus $\alpha_1 + \beta_1 = n$ and the theorem is proved.

COROLLARY 2.3. *If G is a finite graph with each node incident with at least one edge, then $\alpha_0 = \beta_1$ if and only if $\alpha_1 = \beta_0$.*

From the preceding theorem $\alpha_0 + \beta_0 = \alpha_1 + \beta_1$. Thus if $\alpha_0 = \beta_1$, $\alpha_1 = \beta_0$ and vice versa. Since according to Theorem 2.1, $\alpha_1 = \beta_0$ for finite bipartite graphs, we conclude that also $\alpha_0 = \beta_1$ and we have the second theorem of König.

THEOREM 2.4. *For a finite bipartite graph with each node incident with at least one edge, the maximum cardinality of a stable set of nodes equals the minimum cardinality of a set of edges that cover all nodes.*

For a graph that is not bipartite, it need not be the case that $\alpha_0 = \beta_1$ and $\alpha_1 = \beta_0$. The simplest example is a graph with 3 nodes x, y, z and edges $\{x, y\}$, $\{y, z\}$, $\{z, x\}$. For this graph $\alpha_0 = 1$, $\beta_1 = 2$, $\alpha_1 = 1$, $\beta_0 = 2$. The non-bipartite graph pictured in Figure 2 has $\alpha_0 = \beta_0 = \alpha_1 = \beta_1 = 2$.

Theorem 2.4 can be used to answer the question concerning the covering of the squares of a "pruned board" with dominos by

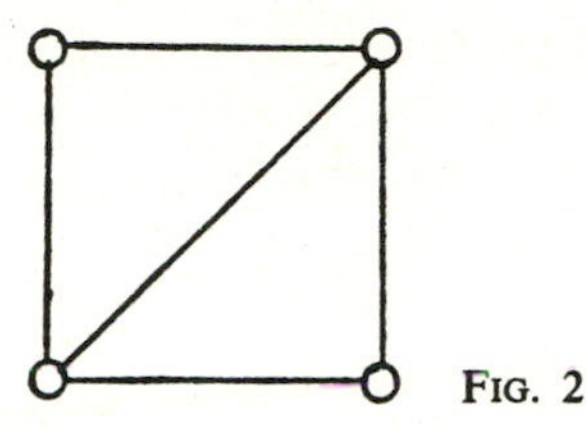

FIG. 2

associating the bipartite graph $\langle R, \Delta, B\rangle$ as before. A stable set of nodes in this bipartite graph corresponds to a set of squares of the "pruned board" no two of which can be simultaneously covered by one domino while a set of edges that cover all nodes corresponds to a set of dominos on the "pruned board" which cover all the squares.

COROLLARY 2.5. *Given a "pruned chessboard" in which each square is adjacent to at least one other, the maximum number of squares no two of which can be simultaneously covered by one domino equals the minimum number of dominos which can be placed on the "pruned chessboard" to cover all squares.*

Most everyone has come across the following problem: If one cuts out two diagonally opposite squares of an ordinary 8×8 chessboard, then it is impossible to cover the resulting "pruned board" with 31 dominos (since there are 62 squares left, at least 31 dominos are necessary). This is so since the 'pruned board' has 30 red squares and 32 black squares and each domino covers one square of each color. Suppose, however, we cut out exactly one red square and exactly one black square, so that the resulting 'pruned board' has 31 red squares and 31 black squares. Is it possible to perfectly cover the squares of the "pruned board" with 31 dominos? Equivalently, does the associated bipartite graph $\langle R, \Delta, B\rangle$ always have a matching of cardinality 31? We shall use Corollary 2.5 to answer this question and the corresponding question for chessboards of any size. Thus we shall not approach the problem from the point of view of matchings, as we have remarked we can, but from the point of view of coverings.

THEOREM 2.6. *Consider an $m \times n$ chessboard whose squares are alternately colored red and black. Suppose at least one of m and n is even and both are greater than one. Then if one red square and one black square are cut out of the board, it is always possible to perfectly cover the resulting 'pruned board' with $(mn-2)/2$ dominos.**

Since either m or n is even, the $m \times n$ chessboard has $mn/2$ red squares and the same number of black squares. Thus after removing one red and one black square, the "pruned board" has an equal number, $(mn-2)/2$, of red and black squares. Let $\langle R, \Delta, B \rangle$ be the bipartite graph associated with the "pruned board." Since m and n are both greater than one, each square of the $m \times n$ board is adjacent to at least two others. Since to get the "pruned board" we have removed only one red and one black square, each square of the "pruned board" is adjacent to at least one other. Thus Corollary 2.5 applies.

Suppose it were impossible to perfectly cover the "pruned board" with $(mn-2)/2$ dominos. Since the "pruned board" can be covered with $(mn-2)/2$ dominos if and only if it can be perfectly covered with $(mn-2)/2$ dominos, then according to Corollary 2.5 there must exist a set S of $mn/2$ squares of the 'pruned board' no two of which are adjacent (a stable set of cardinality $mn/2$ of the associated bipartite graph). We shall have arrived at a contradiction when we show that in an $m \times n$ chessboard ($m, n > 1$, m or n even) a collection of $mn/2$ squares no two of which are adjacent consists either of all the red squares or all the black squares.

For definiteness, suppose n is even. In any row of the $m \times n$ board, we can choose at most $n/2$ squares no two adjacent (since $n/2$ dominos cover the whole row of squares). Thus the set S contains exactly $n/2$ squares from each row. If S contains all

*Since writing this article, we have seen an elegant direct proof of this theorem by Ralph Gomory; see *Mathematical Gems* by Ross Honsberger (The Mathematical Association of America, 1973, pages 66–67).

squares of one color, say red, in some row, then S must also contain all the red squares in the row immediately before and after this row and, by repeating this argument, S contains only red squares. So suppose S contains squares of both colors from each row, in particular from the first row. Since S must contain either the first or last square of each row (the interior squares can be covered with $(n/2) - 1$ dominos), we may assume S contains the first square of row 1 (the corner square) and that it is red. Suppose the first black square in S is the tth square of the row (t is even), and consider the second row. S must contain $n/2$ squares of the second row and S can contain neither the first nor tth square of the row. Since the remaining squares can be covered by

$$\frac{t-2}{2} + \frac{n-t}{2} = \frac{n}{2} - 1$$

dominos, S can contain at most $(n/2) - 1$ squares from the second row. This is a contradiction, so that S consists of the red squares or only of the black squares, and as we have remarked, this proves the theorem.

An $m \times n$ chessboard with both m and n odd has $(mn + 1)/2$ squares of one color (let us fix this to be the red color, so that all corner squares are colored red) and $(mn - 1)/2$ squares of the other color. Since there is one more red square than black square, it is impossible to perfectly cover the squares of the chessboard with dominos. But if we cut out a red square, the preceding discrepancy disappears. We show in the next theorem that we can even do better.

THEOREM 2.7. *Consider an $m \times n$ chessboard whose squares are alternately colored red and black with m and n both odd integers greater than one. Then if two red squares and one black square are cut from the board, it is always possible to perfectly cover the resulting 'pruned board' with $(mn - 3)/2$ dominos.*

The method of proof is the same as that of Theorem 2.6. Every square of the $m \times n$ board, except for the four corner red squares are adjacent to 3 or 4 other squares. The four corner red squares are adjacent to two black squares. Since the "pruned board" results by cutting out two red and one black square, each of its squares is adjacent to at least one other. According to Corollary 2.5 then, if we cannot perfectly cover the "pruned board" with $(mn - 3)/2$ dominos, there exists a set S of $(mn - 1)/2$ squares of the "pruned board" no two of which are adjacent.

S contains at most $(n + 1)/2$ squares from each row. Moreover, if S does contain $(n + 1)/2$ squares from a row, they must be the first, third, . . . squares in that row (and thus are all of the same color). For, if S does not contain the first square of the row, the remaining squares can be covered with $(n - 1)/2$ dominos and thus S can contain at most $(n - 1)/2$ squares of the row. A simple induction completes the verification of the above statement. For at least $(m - 1)/2$ rows it must be true that S contains $(n + 1)/2$ squares of the row. Otherwise S contains at most

$$\left(\frac{m-3}{2}\right)\left(\frac{n+1}{2}\right)+\left(\frac{m+3}{2}\right)\left(\frac{n-1}{2}\right)=\frac{mn-3}{2}$$

squares, which is a contradiction. No row of which S contains $(n + 1)/2$ squares can follow another. Thus either S contains the $(n + 1)/2$ red squares of each of rows $1, 3, 5, \ldots, m$ or else S contains $(n + 1)/2$ squares of each of $(m - 1)/2$ rows no two of which are consecutive. In the latter case, all but at most one of the other rows precedes or follows one of these $(m - 1)/2$ rows.

Suppose it is the case that S contains all the red squares in rows $1, 3, 5, \ldots, m$. Since every black square in the remaining rows is adjacent to one of these red squares, S must contain only red squares. But the "pruned board" has only $(mn - 3)/2$ red squares, and this is a contradiction.

If it is the case that S contains the $(n + 1)/2$ squares of one color in each of $(m - 1)/2$ rows, then S must contain $(n - 1)/2$ squares from the remaining rows, since

$$\left(\frac{m-1}{2}\right)\left(\frac{n+1}{2}\right)+\left(\frac{m+1}{2}\right)\left(\frac{n-1}{2}\right)=\frac{mn-1}{2}.$$

Let these $(m - 1)/2 = k$ rows be rows $i_1, i_2, \ldots, i_k$ where $1 \leqslant i_1 < \ldots < i_k \leqslant m$. If $i_1, \ldots, i_k$ all have the same parity then the

$k(n + 1)/2$ squares of S that come from these rows are all of the same color, say red. But then the squares of S in any row next to one of these rows must consist of all the red squares of that row. This accounts for all squares in S except possibly for those coming from one row. But all black squares of this remaining row are adjacent to the red squares of any row that follows or precedes it. Hence the squares of this row that are in S are red too. Thus S consists only of red squares. On the other hand if $i_1, \ldots, i_k$ are not all of the same parity choose the first consecutive pair i_j, i_{j+1} which are of different parity. Then $i_{j+1} - i_j - 1$ must be 2. For if not, being even it would be at least 4, and there would be two or more rows not preceding or following one of rows $i_1, \ldots, i_k$. But, if, say, i_j is odd and i_{j+1} is even, then S contains all the red squares in row i_j and all the black squares in row i_{j+1}. Thus S must contain all the red squares in row $i_j + 1$ and all the black squares in row $i_{j+1} - 1 = i_j + 2$. But each such red square is adjacent to one of these black squares and we have a contradiction.

We conclude that any set S of $(mn - 1)/2$ squares of the $m \times n$ chessboard no two of which are adjacent all have the same color. Thus the "pruned board", having only $(mn - 3)/2$ squares of each color, cannot contain a set of $(mn - 1)/2$ squares no two of which are adjacent. As we have seen, this proves the theorem.

3. A TIME-TABLE PROBLEM

We take up now the third problem posed in the introduction. This is the problem of designing a time-table requiring the fewest number of hours. There are m workers $w_1, \ldots, w_m$ and a gadget to be made which is broken down into n units $u_1, \ldots, u_n$. Worker w_i is required to spend a_{ij} (a whole number) hours to complete his or her task on unit u_j $(1 \leqslant i \leqslant m,\ 1 \leqslant j \leqslant n)$. We construct a *bipartite multi-graph* $\langle W, \Delta, U\rangle$, that is, a bipartite graph in which several edges may be incident with the same two nodes. For this multigraph, $W = \{w_1, \ldots, w_m\}$, $U = \{u_1, \ldots, u_n\}$, and there are a_{ij} edges in Δ joining w_i and u_j. Now suppose Δ can be partitioned into t matchings $\Delta_1, \ldots, \Delta_t$. We can regard Δ_k $(1 \leqslant k \leqslant t)$ as the schedule for the kth hour by agreeing that if there is an edge in Δ_k joining w_i and u_j then w_i spends hour k on unit u_j. Since Δ_k is a matching, no worker is required to spend the kth hour on two different units nor during the kth hour is any unit receiving the attention of more than one worker. Thus the partition $\Delta_1, \ldots, \Delta_t$ furnishes a timetable which permits one of the gadgets to be made in t hours.

Now as remarked in the preface if $r_i = \Sigma_{j=1}^{n} a_{ij}$ $(1 \leqslant i \leqslant m)$, $s_j = \Sigma_{i=1}^{m} a_{ij}$ $(1 \leqslant j \leqslant n)$, and $p = \max\{r_1, \ldots, r_m, s_1, \ldots, s_n\}$, then any timetable will require at least p hours. The question is whether p hours suffice. If we define the *degree of a node x*, deg x, of a multi-graph to be the number of edges incident with the node x, then in $\langle W, \Delta, U \rangle$ deg $w_i = r_i$ $(1 \leqslant i \leqslant m)$, deg $u_j = s_j$ $(1 \leqslant j \leqslant n)$, and p is the maximal degree of a node. Hence if we can show that it is always possible to partition the edges of a bipartite multi-graph into p matchings where p is the maximal degree of a node, then p hours will always suffice for a timetable. We shall prove this, taking Theorem 2.1 as our principal tool. We first prove the following theorem:

THEOREM 3.1. *Let $\langle X, \Delta, Y \rangle$ be a finite bipartite multi-graph in which the maximal degree of a node is p. Let X_1 be the set of nodes in X with degree equal to p. Then there exists a matching M which matches X_1 to a subset of Y.*

Consider the multi-graph $\langle X_1, \Delta_1, Y \rangle$ where Δ_1 consists of all those edges in Δ which have one of their nodes in X_1. Thus the degree of each node in X_1 with respect to this multi-graph is p and no node in Y has degree greater than p. We need to show that there is a matching M in $\langle X_1, \Delta_1, Y \rangle$ of cardinality $|X_1|$. According to Theorem 2.1* such a matching M exists if and only if this multi-graph has no separating set of cardinality less than $|X_1|$.

*Theorem 2.1 was stated and proved for bipartite graphs but applies equally well to bipartite multi-graphs. Indeed neither the maximum nor the minimum in the theorem can change if several edges now join two nodes when originally only one did.

Suppose $A \cup B$ were a separating set with $A \subseteq X_1$, $B \subseteq Y$, $|A \cup B| \leqslant |X_1| - 1$. Then every edge is incident with at least one node in $A \cup B$. Since the degree of any node is at most p, there are at most $p(|X_1| - 1)$ edges in Δ_1. But since the degree of each node in X_1 is p, there are exactly $p|X_1|$ edges in Δ_1. This is a contradiction, which proves that the matching M exists.

According to Theorem 3.1, if $\langle X, \Delta, Y \rangle$ is a finite bipartite multigraph for which the maximal degree of a node is p and X_1, respectively Y_2, is the set of nodes of X, respectively Y, which have degree equal to p, then there exist matchings Δ_1 and Δ_2 with Δ_1 matching X_1 with $Y_1 \subseteq Y$ and Δ_2 matching Y_2 with $X_2 \subseteq X$. The next theorem and its proof, due to Mendelsohn and Dulmage [21], shows how to find within $\Delta_1 \cup \Delta_2$ a matching which both X_1 and Y_2 meet.

THEOREM 3.2. *Let $\langle X, \Delta, Y \rangle$ be a finite bipartite multi-graph and let Δ_i be a matching which matches $X_i \subseteq X$ with $Y_i \subseteq Y (i = 1, 2)$. Then there is a matching $\Delta' \subseteq \Delta_1 \cup \Delta_2$ which matches $X' \subseteq X$ with $Y' \subseteq Y$ where $X_1 \subseteq X'$ and $Y_2 \subseteq Y'$.*

To prove this theorem, we shall use the much applied method of alternating paths. Consider the bipartite graph $\langle X_1 \cup X_2, \Delta_1 \cup \Delta_2, Y_1 \cup Y_2 \rangle$. Each node of this graph has degree 1 or 2; hence the connected components of this graph are either paths or cycles whose edges alternate being in Δ_1 and Δ_2. Each node $y \in Y_2 \setminus Y_1$, being incident only with an edge in Δ_2, is in a connected component which is a path P_y joining y either to a node $x \in X_2 \setminus X_1$ or a node $z \in Y_1 \setminus Y_2$. In the former case, the last edge of the path is in Δ_2; in the latter case it is in Δ_1. Let Δ_1^y consist of those edges in Δ_1 and Δ_2^y the edges in Δ_2 which are edges of one of the paths $P_y (y \in Y_2 \setminus Y_1)$. Then $(\Delta_1 \setminus \Delta_1^y) \cup \Delta_2^y$ is a matching. If P_y joins y to x, it matches $X_1 \cup \{x\}$ with $Y_2 \cup \{y\}$. If P_y joins y to z, then it matches X_1 with $(Y_1 \setminus \{z\}) \cup \{y\}$. The set of edges

$$\left(\Delta_1 \setminus \bigcup_{y \in Y_2 \setminus Y_1} \Delta_1^y\right) \cup \bigcup_{y \in Y_2 \setminus Y_1} \Delta_2^y$$

is a matching having the required properties.

Combining Theorems 3.1 and 3.2 as already indicated gives the following theorem:

THEOREM 3.3 *Let $\langle S, \Delta, Y\rangle$ be a finite bipartite multi-graph with the maximal degree of a node equal to p. Then there exists a matching Δ' which matches $X' \subseteq X$ with $Y' \subseteq Y$ where X' and Y' contain all nodes of degree equal to p.*

The final theorem of this section furnishes the sought after solution of the timetable problem.

THEOREM 3.4. (König [19]. *Let $\langle X, \Delta, Y\rangle$ be a finite bipartite multi-graph with the maximal degree of a node equal to p. Then the set of edges Δ can be partitioned into p matchings $\Delta_1, \ldots, \Delta_p$.*

This theorem follows readily by induction from Theorem 3.3. If $p = 1$, Δ itself is a matching. If $p > 1$, then by Theorem 3.3 there exists a matching Δ_1 such that the maximal degree of a node of the multi-graph $\langle X, \Delta\backslash\Delta_1, Y\rangle$ is $p - 1$. By induction $\Delta\backslash\Delta_1$ can be partitioned into $p - 1$ matchings $\Delta_2, \ldots, \Delta_p$. Then $\Delta_1, \Delta_2, \ldots, \Delta_p$ is a partition of Δ into p matchings.

The method of alternating paths which yielded Theorem 3.2 will be applied again in sections 4 and 6 to both finite and infinite graphs.

4. TRANSFINITE FAMILIES OF SETS

To introduce this section, we return to the chessboard which we now assume extends infinitely far in all directions (formally, we consider the regular tessellation of the plane by squares). We assume as before that the squares are colored alternately red and black. We take a pair of scissors and cut out as many squares as we wish (possibly an infinite number) to arrive at a "pruned board." When is it possible to perfectly cover the "pruned board" with dominos? As in the finite case, we can associate a bipartite graph $\langle R, \Delta, B\rangle$ with the "pruned board," the only difference being that now the sets of nodes R and B may be infinite. While this is the case, every node of the graph is incident with only a finite number of edges (indeed at most 4 edges). Thus if for $r \in R$, B_r is the set of all black squares of the "pruned board" whicn are adjacent to the red square r and if for $b \in B$, R_b is the set of all red squares of the "pruned board" adjacent to the black square b, then both $(B_r : r \in R$ and $(R_b : b \in B)$ are (possibly infinite) families of finite sets.

The problems that arise when R and B are allowed to be infinite sets are significant. For a finite "pruned board" we concluded that a perfect covering of the squares exists if and only if the sets R and B have the same finite cardinal number and there is a matching in $\langle R, \Delta, B\rangle$ with cardinality equal to $|R|$. This type of reasoning breaks down when R and B are infinite. Thus it is possible that R and B have the same infinite cardinal number, that there is a matching in $\langle R, \Delta, B\rangle$ with R meeting the matching but no matching exists which both R and B meet. The "pruned board" in Figure 3 is an example where it is trivial to place dominos in a non-overlapping fashion with all red squares covered but where it is impossible to place dominos in a non-overlapping fashion with all the squares (red and black) covered. Yet R and B are both countably infinite.

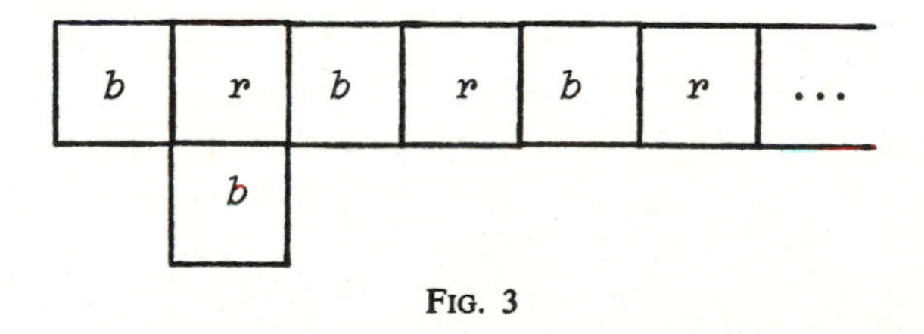

FIG. 3

But suppose there is a matching in $\langle R, \Delta, B\rangle$ which matches R to a subset of B and another matching which matches B to a subset of R. Is there then a matching which matches R to B? In other words, if it is possible to place dominos in one way so that all red squares are covered and in another way so that all black squares are covered, is it possible to place dominos so that both the red and black squares are covered? We shall prove a theorem in this section which answers this question affirmatively.

We want first to obtain the analog of Theorem 1.1 for infinite families of sets. To do this we will have to assume that the individual sets are finite, as they are in the families $(B_r : r \in R)$ and $(R_b : b \in B)$ above. An important tool for obtaining transfinite analogs of many finite theorems is a selection principle of Rado [28]. This principle gives the existence of a choice function for a family $\mathfrak{A}$ of sets which mimics the behavior of choice functions of finite subfamilies. Suppose $\mathfrak{A} = (A_i : i \in I)$ is a family of subsets of E. Then a choice function θ of $\mathfrak{A}$ is a map $\theta : I \to \cup_{i \in I} A_i$ such that $\theta(i) \in A_i (i \in I)$. Thus $(\theta(i) : i \in I)$ is what we have previously called a system of representatives of $\mathfrak{A}$. If $\langle I, \Delta, E \rangle$ is the bipartite graph associated with $\mathfrak{A}$, then a choice function of $\mathfrak{A}$ is equivalent to a set $\Delta' \subseteq \Delta$, such that each node in I is incident with exactly one edge in Δ'.

For K a finite subset of I, we write $K \subset\subset I$.

THEOREM 4.1. *Let $\mathfrak{A} = (A_i : i \in I)$ be a family of finite subsets of a set E. Suppose for each $K \subset\subset I$ there is given a choice function θ_K of $(A_i : i \in K)$. Then there is a choice function θ of $\mathfrak{A}$ such that given any $J \subset\subset I$ there is K with $J \subseteq K \subset\subset I$ and $\theta(i) = \theta_K(i)$ for $i \in J$.*

Recently Rado [29] gave a short proof of this theorem which we now present. It will be convenient to prove the theorem in the terminology of bipartite graphs. Thus we have a bipartite graph $\langle I, \Delta, E \rangle$ with each node of I incident with only finitely many edges in Δ. For each $K \subset\subset I$ we are given a set $\Delta_K \subseteq \Delta$ with each node of K incident with exactly one edge of Δ_K. We need to show there exists $\Delta^* \subseteq \Delta$ with each node of I incident with exactly one edge of Δ^* such that given $J \subset\subset I$ there is K with $J \subseteq K \subset\subset I$ and $\Delta^* \cap (J \times E) = \Delta_K \cap (J \times E)$†.

---

†$J \times E$ is the set of all *unordered* pairs $\{j, e\}$ with $j \in J$, $e \in E$.

If I is finite we can put $\Delta^* = \Delta_I$. So let I be infinite. Let Ω be the set of all subsets Δ' of Δ such that given any $J \subset\subset I$ there is an L with $J \subseteq L \subset\subset I$ and $\Delta_L \cap (J \times E) \subseteq \Delta'$. By putting $L = J$ we see $\Delta \in \Omega$, so that Ω is non-empty. We partially order Ω by set containment. Suppose $\Delta^1, \Delta^2, \Delta^3, \ldots,$ are in Ω with

$$\Delta^1 \supseteq \Delta^2 \supseteq \Delta^3 \ldots .$$

Let $\Delta^\infty = \cap_{i=1}^\infty \Delta^i$. Take $J \subset\subset I$; since each node of I is incident with only finitely many edges in Δ, there exists an integer k (depending on J) such that $\Delta^\infty \cap (J \times E) = \Delta^k$. If L is such that $J \subseteq L \subset\subset I$ and $\Delta_L \cap (J \times E) \subseteq \Delta^k$, then $\Delta_L \cap (J \times E) \subseteq \Delta^\infty$. Thus $\Delta^\infty \in \Omega$. According to Zorn's lemma there exists a $\Delta^* \in \Omega$ which is minimal with respect to the partial order. We show Δ^* has the required properties.

Let $\{i, e\}$ be an edge† in Δ^*. By the minimality of Δ^* there is a J_e with $i \in J_e \subset\subset I$ such that whenever $J_e \subseteq L \subset\subset I$ and $\Delta_L \cap (J_e \times E) \subseteq \Delta^*$, then $\{i, e\} \in \Delta_L$. Suppose $\{i, e_1\}$ and $\{i, e_2\}$ are in Δ^*. If $J^* = J_{e_1} \cup J_{e_2}$, then $J^* \subset\subset I$ so that there is L^* with $J^* \subseteq L^* \subset\subset I$ and $\Delta_{L^*} \cap (J^* \times E) \subseteq \Delta^*$. By definition of J_{e_1} and J_{e_2}, $\{i, e_1\}, \{i, e_2\} \in \Delta_{L^*}$ so that $e_1 = e_2$. We conclude that each node of I is incident with exactly one edge in Δ^*. Moreover, given $J \subset\subset I$, there is L with $J \subseteq L \subset\subset I$ and $\Delta_L \cap (J \times E) \subseteq \Delta^*$; hence $\Delta_L \cap (J \times E) = \Delta^* \cap (J \times E)$, since each node of I is incident with exactly one edge of Δ^*. This completes the proof of the theorem.

We can now give the transfinite analogue of Theorem 1.1, which was first proved by M. Hall Jr. [15].

THEOREM 4.2. *Let $\mathfrak{A} = (A_i : i \in I)$ be a family of finite subsets of a set E. Then the following are equivalent*:

(1) *$\mathfrak{A}$ has a transversal.*

(2) *Every finite subfamily of $\mathfrak{A}$ has a transversal.*

(3) $|\cup_{i \in J} A_i| \geqslant |J| \qquad (J \subset\subset I)$.

†By definition of Ω, i is incident with some edge in Δ^*.

Surely (1) implies (2), and using Theorem 1.1 we conclude that (2) and (3) are equivalent. Suppose now (2) holds. Then for every $K \subset\subset I$ there is an injective choice function θ_K of the family $(A_i : i \in K)$. Let θ be the choice function of $\mathfrak{A}$ whose existence is given by Theorem 4.1. Let $i_1, i_2 \in I$ with $i_1 \neq i_2$ and set $K = \{i_1, i_2\}$. Then there exists $J \subset\subset I$ with $K \subseteq J$ such that $\theta(i_1) = \theta_J(i_1)$ and $\theta(i_2) = \theta_J(i_2)$. Since θ_J is injective, $\theta(i_1) \neq \theta(i_2)$. Hence θ is an injective choice function of $\mathfrak{A}$, so that $(\theta(i) : i \in I)$ is a system of distinct representatives and $\{\theta(i) : i \in I\}$ a transversal of $\mathfrak{A}$.

It is to be observed that condition (3) of Theorem 4.2 is no longer sufficient for a family of sets to have a transversal if the assumption that the sets are finite is eliminated. The family $(A_i : i = 1, 2, \ldots)$ where $A_1 = \{1, 2, \ldots\}$, $A_2 = \{1\}$, $A_3 = \{2\}, \ldots, A_k = \{k-1\}, \ldots$ satisfies (3) but yet has no transversal.

Theorem 4.2 is equivalent to the following theorem about bipartite graphs:

THEOREM 4.3. *Let $\langle X, \Delta, Y\rangle$ be a bipartite graph in which the degree of each node in X is finite. Then the following are equivalent*:

(1) *X is matched with a subset of Y.*
(2) *Every finite subset of X is matched to a subset of Y.*
(3) *There is no separating set of the form $(X \setminus A) \cup B$ where $A \subset\subset X$, $B \subset\subset Y$ and $|B| < |A|$.*

We now derive the analog of Theorem 3.2 for infinite bipartite graphs. The statement of the theorem remains the same, but for the convenience of the reader we state it again. The theorem is due to Ore [25]; a more exact statement of the situation is derived by Perfect and Pym [26].

THEOREM 4.4. *Let $\langle X, \Delta, Y\rangle$ be a bipartite graph and let Δ_i be a matching which matches $X_i \subseteq X$ with $Y_i \subseteq Y$ $(i = 1, 2)$. Then there is a matching $\Delta' \subseteq \Delta_1 \cup \Delta_2$ which matches $X' \subseteq X$ with $Y' \subseteq Y$ where $X_1 \subseteq X'$ and $Y_2 \subseteq Y'$.*

As a corollary there is the following theorem of Banach [1].

COROLLARY 4.5. *Let* $\langle X, \Delta, Y\rangle$ *be a bipartite graph. Suppose there is a matching* Δ_1 *which matches* X *with a subset of* Y *and a matching* Δ_2 *which matches* Y *with a subset of* X. *Then there is a matching* $\Delta' \subseteq \Delta_1 \cup \Delta_2$ *which matches* X *and* Y.

From Corollary 4.5 one easily deduces the Schroeder-Bernstein theorem that if α and β are two infinite cardinal numbers with $\alpha \leqslant \beta$ and $\beta \leqslant \alpha$ then $\alpha = \beta$.

Notice that Corollary 4.5 answers affirmatively one of the questions that was raised about "pruned boards." Namely, if it is possible to place dominos in one way so that all red squares are covered and in another way so that all black squares are covered, then it is possible to place dominos so that both the black and red squares are covered. Moreover, the positions of these dominos is a subset of the union of the positions of the first two placements.

We now turn to the proof of Theorem 4.4 and consider the bipartite graph $\langle X_1 \cup X_2, \Delta_1 \cup \Delta_2, Y_1 \cup Y_2\rangle$. The proof is essentially that of Theorem 3.2. The only difference is that since the above bipartite graph may be infinite, there are two more possibilities for its connected components. Indeed since every node is incident with 1 or 2 edges, the connected components are either paths (finite, infinite in one direction, or infinite in two directions) or cycles, in all cases the edges alternating in Δ_1 and Δ_2. Each node $y \in Y_2 \setminus Y_1$ is in a connected component which is a path joining y to a node in $Y_1 \setminus Y_2$ or a node in $X_2 \setminus X_1$, or else an infinite path beginning at y. If we let Δ_1^y be the edges of Δ_1 and Δ_2^y the edges of Δ_2 in this connected component, then

$$\left(\Delta_1 \setminus \bigcup_{y \in Y_2 \setminus Y_1} \Delta_1^y\right) \cup \bigcup_{y \in Y_2 \setminus Y_1} \Delta_2^y$$

is a matching with the required properties.

If Theorem 4.3 and Corollary 4.5 are combined, the following theorem is obtained:

THEOREM 4.6. *Let $\langle X, \Delta, Y\rangle$ be a bipartite graph in which the degree of every node is finite. Then there is a matching which meets all nodes of the graph if and only if*

(1) *there is no separating set of the form $(X\backslash A)\cup B$ where $A\subset\subset X$, $B\subset\subset Y$ and $|B|<|A|$.*

(2) *there is no separating set of the form $(Y\backslash C)\cup D$ where $C\subset\subset Y$, $D\subset\subset X$ and $|D|<|C|$.*

Indeed, according to Theorem 4.3, (1) is equivalent to the existence of a matching Δ_1 which matches X with a subset of Y and (2) is equivalent to the existence of a matching which matches Y with a subset of X. According to Corollary 4.5 the existence of such matchings Δ_1 and Δ_2 is equivalent to there being a matching which matches X with Y.

Notice that Theorem 4.6 furnishes necessary and sufficient conditions in order that a "pruned board" can be covered perfectly with dominos. Theorem 4.6 is equivalent to the following theorem about a family of sets. Recall that if $\mathfrak{A}=(A_i : i\in I)$ is a family of subsets of a set E, then $\mathfrak{A}^*=(A_e^* : e\in E)$, where for $e\in E$, $A_e^*=\{i\in I : e\in A_i\}$, is the dual family of $\mathfrak{A}$.

THEOREM 4.7. *Let $\mathfrak{A}=(A_i : i\in I)$ be a family of finite subsets of a set E such that every element of E is a member of only finitely many of the $A_i (i\in I)$. Then E is a transversal of $\mathfrak{A}$ if and only if*

(1) $|\cup_{i\in J}A_i|\geqslant |J| \qquad (J\subset\subset I)$.

(2) $|\cup_{e\in F}A_e^*|\geqslant |F| \qquad (F\subset\subset E)$.

We now take a global point of view. We take a bipartite graph $\langle X, \Delta, Y\rangle$ and let M_X (respectively, M_Y) be the collection of all subsets of X (respectively, Y) which are matched with some subset of Y. It is the nature of the set M_X in which we are now interested. An elementary property is that every subset of a member of M_X is also a member of M_X. If we regard M_X as being partially ordered by set inclusion, then M_X need not contain maximal members. The simplest way to see this is to take X to be an uncountably infinite set and Y a countably infinite set and to put an edge joining each node of X to every node of Y. When the degrees of the nodes in X are finite, then M_X will have a maximal member.

THEOREM 4.8. *Let $\langle X, \Delta, Y\rangle$ be a bipartite graph in which the nodes in X have finite degree. Then M_X contains a maximal set.*

According to Theorem 4.3 a set A is in M_X if and only if every finite subset of A is. Let $A_1 \subseteq A_2 \subseteq A_3 \subseteq \ldots$ be an infinite ascending chain of sets in M_X. Then $A = \cup_{i=1}^{\infty} A_i$ is in M_X, for if $A' \subset\subset A$, then $A' \subseteq A_k$ for some k and hence $A' \in M_X$. If we now invoke Zorn's lemma, we conclude that M_X contains at least one maximal set.

The collection M_X may contain maximal sets even when some of the nodes of X are incident with infinitely many edges. A trivial example is obtained by taking X and Y to be two infinite sets of the same cardinality and all possible edges between X and Y. Since then $X \in M_X$, X is a maximal set in M_X. A less trivial example is one we have met previously as a family of sets. Namely, take $X = Y = \{1, 2, 3, \ldots\}$ and take $\Delta = \{\{1, 1\}, \{1, 2\}, \{1, 3\}, \ldots, \{2, 1\}, \{3, 2\}, \{4, 3\}, \ldots\}$. Then for each positive integer k, $X \setminus \{k\}$ is a maximal set in M_X.

When M_X has maximal members, the following theorem (taken from [6]) gives some information about M_X and the matchings in $\langle X, \Delta, Y\rangle$.

THEOREM 4.9. *Let $\langle X, \Delta, Y\rangle$ be a bipartite graph. Then the following statements hold*:

(1) *If B is a maximal set in M_X and $C \in M_Y$ then there is a matching which matches B and a subset of Y containing C.*

(2) *Let B_1, B_2 be maximal sets in M_X. Let Δ_i be a matching which matches B_i with the set $C_i \subseteq Y$ $(i = 1, 2)$. Then there is a matching $\Delta' \subseteq \Delta_1 \cup \Delta_2$ which matches B_2 with C_1.*

(3) *All maximal sets in M_X have the same cardinality.*

(4) *If B_1 is a maximal set in M_X and $B_2 \in M_X$, then there exists an injection $\sigma : B_2 \setminus B_1 \to B_1 \setminus B_2$ such that $(B_2 \setminus \{\sigma(x)\}) \cup \{x\}$ is*

a maximal set in M_X for each $x \in B_2 \setminus B_1$ and $(B_2 \setminus \{x\}) \cup \{\sigma(x)\} \in M_X$. If B_2 is also a maximal set in M_X, then so is $(B_2 \setminus \{x\}) \cup \{\sigma(x)\}$.

Statement (1) follows easily from Theorem 4.4, for from this theorem we deduce the existence of sets B', C' with $B \subseteq B' \subseteq X$, $C \subseteq C' \subseteq Y$ and a matching which matches B' and C'. Since B is a maximal set in M_X, $B' = B$, which gives the desired conclusion.

To prove (2), we look at the bipartite graph $\langle B_1 \cup B_2, \Delta_1 \cup \Delta_2, C_1 \cup C_2 \rangle$. It follows as in the proof of Theorem 4.4 that the connected components of this graph are either paths (finite, infinite in one direction, infinite in two directions) or cycles, in all cases the edges alternating in Δ_1 and Δ_2. Each node $y_1 \in C_1 \setminus C_2$ is in a connected component which is either a path $P(y_1, y_2)$ joining y_1 to a node $y_2 \in C_2$ or else an infinite path $P(y_1, \cdot)$ beginning at y_1. The only other possibility is a path joining y_1 to a node $x_1 \in B_1 \setminus B_2$; but then if $\Delta_1^{y_1}$ were the edges of Δ_1 and $\Delta_2^{y_1}$ the edges of Δ_2 in this connected component, $(\Delta_2 \setminus \Delta_2^{y_1}) \cup \Delta_1^{y_1}$ would be a matching which matches $B_2 \cup x_1$ with a subset of Y, which contradicts the assumption that B_2 is a maximal set in M_X. Thus the possibilities for the connected component containing the node y_1 are as given. If we use the above notation, then $(\Delta_2 \setminus \Delta_2^{y_1}) \cup \Delta_1^{y_1}$ is a matching which matches B_2 with $(C_2 \setminus \{y_2\}) \cup \{y_1\}$ in the case of the path $P(y_1, y_2)$ and with $C_2 \cup \{y_1\}$ in the case of the path $P(y_1, \cdot)$. At this point we could say, since these paths are connected components and have neither nodes nor edges in common, that

$$\Delta' = \left(\Delta_2 \setminus \bigcup_{y_1 \in Y_1 \setminus Y_2} \Delta_2^{y_1}\right) \cup \bigcup_{y_1 \in Y_1 \setminus Y_2} \Delta_1^{y_1}$$

is a matching which matches B_2 with a subset of Y containing C_1. But we want no nodes outside of C_1 to meet the required matching. We therefore must look at the nodes in $C_2 \setminus C_1$. By symmetry, each node $z_2 \in C_2 \setminus C_1$ is in a connected component which is either a path $Q(z_2, z_1)$ joining z_2 to a node $z_1 \in C_1 \setminus C_2$ or else an infinite path beginning at z_1. In the first case $Q(z_2, z_1) = P(z_1, z_2)$ and the node z_2 does not meet the matching Δ'. It is precisely the nodes z_2 of the second case that must be eliminated. We do this as follows. Let Δ_1^* (respectively, Δ_2^*) consist of all those edges of Δ_1 (respectively, Δ_2) which are in the connected components $P(y_1, y_2)$, $P(y_1, \cdot)$, $Q(z_2, z_1)$, $Q(z_2, \cdot)$; then

$$(\Delta_2 \backslash \Delta_2^*) \cup \Delta_1^*$$

is the required matching.

Statement (3) is an immediate consequence of (2).

The proof of (4) uses the same technique to prove (2). We consider the bipartite graph $\langle B_1 \cup B_2, \Delta_1 \cup \Delta_2, C_1 \cup C_2 \rangle$ where Δ_i matches $B_i \subseteq X$ with $C_i \subseteq Y$ ($i = 1, 2$), and B_1 is a maximal set in M_X. Each node $x \in B_2 \backslash B_1$ lies in a connected component which, because B_1 is a maximal set in M_X, is a path $P(x, \sigma(x))$ joining x to a node $\sigma(x) \in B_1 \backslash B_2$. The map $\sigma : B_2 \backslash B_1 \rightarrow B_1 \backslash B_2$ so defined is an injection, since these paths are connected components and have neither nodes nor edges in common. If Δ_1^x and Δ_2^x are, respectively, the edges in Δ_1 and Δ_2 of the path $P(x, \sigma(x))$ ($x \in B_2 \backslash B_1$), then

$$\Delta_1^* = (\Delta_1 \backslash \Delta_1^x) \cup \Delta_2^x, \qquad \Delta_2^* = (\Delta_2 \backslash \Delta_2^x) \cup \Delta_1^x$$

are matchings which match $(B_1 \backslash \{\sigma(x)\}) \cup \{x\}$ to a subset of Y and $(B_2 \backslash \{x\}) \cup \{\sigma(x)\}$ to a subset of Y. Suppose $B_1' = (B_1 \backslash \{\sigma(x)\}) \cup \{x\}$ were not a maximal set in M_X. Then there exists $z \notin B_1'$ with $B_1' \cup \{z\} \in M_X$. If we replace B_2 with $B_1' \cup \{z\}$ in the preceding argument, we conclude that there is an injection τ from $(B_1' \cup \{z\}) \backslash B_1 = \{x, z\}$ to $B_1 \backslash (B_1' \cup \{z\}) = \{\sigma(x)\}$. Since $z \neq x$, this is a contradiction. Hence $(B_1 \backslash \{\sigma(x)\}) \cup \{x\}$ is a maximal set in M_X.

A similar argument shows that if B_2 is a maximal set in M_X so is $(B_2 \backslash \{x\}) \cup \{\sigma(x)\}$.

To conclude this section we take a look at what Theorem 4.9 means in terms of families of sets.

Suppose $\mathfrak{A} = (A_i : i \in I)$ is a family of subsets of a set E. We form the bipartite graph $\langle I, \Delta, E \rangle$ as before, so that M_I consists of those subsets J of I such that the subfamily $(A_i : i \in J)$ has a transversal while M_E is the collection of partial transversals of $\mathfrak{A}$. A maximal member of M_I is a subset J of I such that $(A_i : i \in J)$ has a transversal but no subfamily $(A_i : i \in K)$ where $J \subsetneq K$ has a transversal. Statement (3) then says that any two subfamilies of $\mathfrak{A}$

which are maximal with respect to the property of having a transversal have the same cardinality. Statements (1) and (2) translate to (1′) and (2′) below.

(1′) *If J is a subset of I such that the subfamily $(A_i : i \in J)$ is maximal with respect to the property of having a transversal, then the collection of partial transversals of $(A_i : i \in I)$ is the same as the collection of partial transversals of $(A_i : i \in J)$.*

(2′) *If J_1, J_2 are two subsets of I such that the subfamilies $(A_i : i \in J_1)$ and $(A_i : i \in J_2)$ are both maximal with respect to the property of possessing a transversal, then each transversal of $(A_i : i \in J_1)$ is a transversal of $(A_i : i \in J_2)$ and vice versa.* (Note that (1′) said only that a transversal of $(A_i : i \in J_1)$ is a partial transversal of $(A_i : i \in J_2)$).

5. MATCHINGS IN GRAPHS

Thus far, we have been primarily concerned with the concept of a matching in a bipartite graph and several of its interpretations. In a finite bipartite graph we have been able to give a criterion for the existence of a matching of a prescribed cardinality. But the notion of a matching applies to graphs which are not bipartite, and one naturally wonders whether the criterion for bipartite graphs can be extended to graphs in general. We shall prove a theorem which is analogous to Theorem 2.1 and then specialize it to obtain a criterion of Berge for the existence of a matching of prescribed cardinality and a major theorem of Tutte for the existence of a matching which all nodes meet, a perfect matching. The theorems can be extended to infinite graphs when nodes have finite degree.

Let G be a graph with N its set of nodes. For $T \subseteq N$, recall that *the subgraph of G induced by T, G_T,* is the graph whose set of nodes is T where two nodes in T are joined by an edge if and only if they are joined in G. If $S \subseteq T \subseteq N$, then clearly $(G_T)_S = G_S$ and we shall make implicit use of this throughout.

Let $T \subseteq N$. The graph $G_{N \setminus T}$ need not be connected even if G is, and thus has in general several connected components. By an *odd component* of $G_{N\setminus T}$ we shall mean a connected component of $G_{N\setminus T}$ which has an odd number of nodes. For A an arbitrary but fixed set of nodes of G, we define $p(T; A)$ to be the number* of odd components of $G_{N\setminus T}$ all of whose nodes belong to A. Thus if $A = N, p(T; N)$ is the number of odd components of $G_{N\setminus T}$.

The proof of the following theorem is a reworking of a proof [5] of a more general theorem which applies to infinite graphs.

THEOREM 5.1. *Let G be a finite graph with node set N. Let $A \subseteq N$ and let t be a positive integer. Then there exists a matching in G which at least t nodes of A meet if and only if*

$$p(T; A) \leqslant |A| + |T| - t \qquad (T \subseteq N). \tag{5.1}$$

Thus the maximum cardinality of a subset of nodes of A which meets a matching equals

$$\min\{|A| + |T| - p(T; A) : T \subseteq N\}.$$

In particular A meets a matching if and only if

$$p(T; A) \leqslant |T| \qquad (T \subseteq N).$$

It should be observed that (5.1) is a criterion for the existence of a matching which t nodes of A meet. The cardinality of this matching may vary due to the fact that some edges of the matching may have both their nodes in A, while others may have only one.

Suppose $B \subseteq A$ with $|B| = t$ and there is a matching Δ which B meets. Let $T \subseteq N$. For each odd component of $G_{N\setminus T}$ all of whose nodes belong to B (and thus to A) there must be an edge in Δ which joins a node of the odd component to T. Thus the number of such odd components does not exceed $|T|$. Hence

$$(|A \cap (N\setminus T)| - p(T; A)) + |T| \geqslant |B \cap (N\setminus T)|.$$

*If there are an infinite number, we take $p(T; A) = \infty$.

Adding $|A \cap T|$ to both sides and using

$$|A \cap T| + |B \cap (N \setminus T)| \geqslant |B| = t,$$

we obtain

$$|A| - p(T; A) + |T| \geqslant t,$$

which is the desired inequality.

We now turn to the sufficiency of (5.1) for the existence of a matching which t nodes of A meet, and to indicate the dependency of $p(T; A)$ on the graph G we write instead $p(G : T; A)$. The proof of sufficiency will be by induction on t. If $t = 1$ and (5.1) is satisfied, then at least one node of A is incident with an edge of G. Otherwise each node of A is an odd connected component of $G = G_{N \setminus \varnothing}$ so that $p(G : \varnothing; A) = |A|$ and (5.1) is violated when $T = \varnothing$. Assume now that $t > 1$. Two cases need to be distinguished.

CASE 1. There is a non-empty subset T of N for which

$$p(G : T; A) = |A| + |T| - t. \tag{5.2}$$

Let $G_{N_k} (k \in K)$ be the connected components of $G_{N \setminus T}$ which either have an even number of nodes or else have an odd number of nodes not all of which belong to A. Set $A_k = N_k \cap A$ $(k \in K)$. Since

$$|A_k| \leqslant |A| - p(G : T; A) = t - |T| \leqslant t - 1,$$

$|A_k| \leqslant t - 1$. Suppose there were an $S \subseteq N_k$ such that

$$p(G_{N_k} : S; A_k) > |S| = |A_k| + |S| - |A_k|.$$

Then

$$\begin{aligned} p(G : T \cup S; A) &= p(G : T; A) + p(G_{N_k} : S; A_k) \\ &< (|A| + |T| - t) + |S| \\ &= |A| + |S \cup T| - t, \end{aligned}$$

and we have contradicted (5.1). Thus for $S \subseteq N_k$, $p(G_{N_k} : S; A_k) \leqslant |S|$, and by the inductive hypothesis there exists a matching Δ_k in G_{N_k} which all nodes of A_k meet. This is true for each $k \in K$.

Now let $G_{T_j}(j \in J)$ be the connected components of $G_{N \setminus T}$ which have an odd number of nodes all of which belong to A. Suppose for some $S \subseteq T_j$ it were true that

$$p(G_{T_j} : S; T_j) > |T_j| + |S| - (|T_j| - 1) = |S| + 1.$$

(Note that $A \cap T_j = T_j$ in this case.) Then a calculation like the preceding one gives

$$p(G_{T_j} : T \cup S; A) > |A| + |T| - t$$

which again contradicts (5.1). Since $|T_j| \leqslant t - 1$, we conclude by the inductive hypothesis that G_{T_j} has a matching which all but one node of T_j meets. This is so for each $j \in J$.

Suppose for some $j \in J$ there were a node $z \in T_j$ such that $G_{T_j \setminus \{z\}}$ did not have a matching meeting all nodes of $T_j \setminus \{z\}$. Then by the inductive hypothesis, there exists $S \subseteq T_j \setminus \{z\}$ such that

$$p\big(G_{T_j \setminus \{z\}} : S; T_j \setminus \{z\}\big) \geqslant |S| + 1.$$

But $|T_j \setminus \{z\}|$ is an even integer so that $p(G_{T_j \setminus \{z\}} : S; T_j \setminus \{z\}) - |S|$ is a positive even integer. Thus we have the stronger inequality

$$p\big(G_{T_j \setminus \{z\}} : S; T_j \setminus \{z\}\big) \geqslant |S| + 2.$$

We now calculate that

$$\begin{aligned} p(G : T \cup S \cup \{z\}; A) &= (p(G : T; A) - 1) \\ &\quad + p\big(G_{T_j \setminus \{z\}} : S; T_j \setminus \{z\}\big) \\ &\geqslant (|A| + |T| - t - 1) + |S| + 2 \\ &= |A| + |T \cup S \cup \{z\}| - t. \end{aligned}$$

This inequality in conjunction with (5.1) (with T replaced by

$T \cup S \cup \{z\}$) means that

$$p(G : T \cup S \cup \{z\}; A) = |A| + |T \cup S \cup \{z\}| - t.$$

This in turn means that in (5.2) we may replace T by $T \cup S \cup \{z\}$.*

Repeating this enlargement of T as much as necessary, we eventually arrive at a set $T \subseteq N$ which satisfies (5.2) such that for each $z \in T_j (j \in J)$ there is a matching $\Gamma_j(z)$ in $G_{T_j \setminus \{z\}}$ which all nodes of $T_j \setminus \{z\}$ meet.

We now construct a bipartite graph $\langle T, \Delta, J \rangle$, where for $t \in T$, $j \in J$, there is an edge joining t and j if and only if there is an edge in G joining t to *some* node in T_j. We assert that this bipartite graph has a matching $\bar{\Delta}$ of cardinality equal to $|T|$. If not, then by Theorem 2.1 it has a separating set $T^0 \cup J^0 (T^0 \subseteq T, J^0 \subseteq J)$ with $|T^0 \cup J^0| < |T|$. We then calculate that

$$\begin{aligned} p(G : T_0; A) \geqslant |J \setminus J_0| = |J| - |J_0| &> |J| + |T_0| - |T| \\ &= p(G : T; A) + |T_0| - |T| \\ &= (|A| + |T| - t) + |T_0| - |T| \\ &= |A| + |T_0| - t. \end{aligned}$$

This contradicts (5.1). Hence $\bar{\Delta}$ exists. Let the edges in $\bar{\Delta}$ be $\{t_1, j_1\}, \ldots, \{t_m, j_m\}$ where $m = |T|$. Choose a node $z_{j_k} \in T_{j_k}$ such that $\{t_k, z_{j_k}\}$ is an edge in G $(k = 1, \ldots, m)$. Let $\bar{\bar{\Delta}}$ be the set of these m edges. For $j \in J \setminus \bar{J}$ where $\bar{J} = \{j_1, \ldots, j_k\}$, choose any node z_j in T_j. Then

$$\left(\bigcup_{k \in K} \Delta_k \right) \cup \left(\bigcup_{j \in J \setminus \bar{J}} \Gamma_j(z_j) \right) \cup \left(\bigcup_{k=1}^{m} \Gamma_{j_k}(z_{j_k}) \right) \cup \bar{\bar{\Delta}}$$

*Thus if we had chosen T to be a maximal subset of N satisfying (5.2), then $S \cup \{z\} \subseteq T$, which is a contradiction. This part of the argument did not use the fact that $T \neq \emptyset$.

is a matching in G which meets $|A| - (p(G : T; A) - |T|) = t$ nodes of A.

CASE 2. For all non-empty subsets T of N,

$$p(G : T; A) < |A| + |T| - t.$$

There are two subcases we wish to consider.

(a) There exists an edge $\{x, y\}$ in G with $x \in A$ and $y \notin A$.

Let $N' = N \setminus \{y\}$, $A' = A \setminus \{x\}$, and $G' = G_{N'}$. Suppose for some $T' \subseteq N'$ that

$$p(G' : T'; A') \geqslant |A'| + |T'| - (t - 2).$$

But $p(G : T' \cup \{y\}; A) = p(G' : T'; A') + 1$ if x is a node of a connected component of $G'_{N' \setminus T'}$ which has an odd number of nodes all of which except x belong to A', while $p(G : T' \cup y;\ A) = p(G' : T';\ A')$ otherwise. Thus $p(G : T' \cup \{y\};\ A) \geqslant p(G' : T'; A')$, so that

$$\begin{aligned} p(G : T' \cup \{y\}; A) &\geqslant |A'| + |T'| - t + 2 \\ &= |A| + |T' \cup \{y\}| - t. \end{aligned}$$

We must have equality throughout and since $T' \cup \{y\} \neq \emptyset$, we contradict the basic assumption of Case 2. Hence

$$p(G' : T'; A') \leqslant |A'| + |T'| - (t - 1) \qquad (T' \subseteq N'),$$

and so by the inductive hypothesis G' has a matching Δ' which at least $t - 1$ nodes of A' meet. If an edge of Δ' is incident with x, then at least t nodes of A meet Δ'; if not, $\Delta' \cup \{x, y\}$ is a matching in G which at least t nodes of A meet.

(b) No node of A is joined by an edge to a node not in A.

We may then assume, clearly, that $N = A$, so that every matching in G meets an even number of nodes in A. If t is not even, we may replace t by $t + 1$. If we are no longer in Case 2, we return to Case 1 where the inductive hypothesis will be applied to an integer smaller than $t + 1$ by an even integer (thus smaller than t). Otherwise we proceed as follows:

Let $\{x,y\}$ be an edge of G. Let $N' = N\setminus\{x,y\}$. Suppose for some $T' \subseteq N'$ that

$$p(G_{N'} : T'; N') > |N'| + |T'| - (t-2).$$

Since $|N'| + |T'| - p(G_{N'} : T'; N')$ is an even integer as t is, we have that

$$\begin{aligned} p(G : T; N) &= p(G_{N'} : T'; N') \\ &\geqslant |N'| + |T'| - (t-4) \\ &= |N| + |T| - t, \end{aligned}$$

where $T = T' \cup \{x,y\}$. We must have equality throughout and since $T \neq \emptyset$, we again contradict the basic assumption of Case 2. Thus

$$p(G_{N'} : T'; N') \leqslant |N'| + |T'| - (t-2) \qquad (T' \subseteq N'),$$

and by the inductive hypothesis G' has a matching Δ' which $t-2$ nodes of N' meet. Thus $\Delta' \cup \{x,y\}$ is a matching of G which t nodes of N $(= A)$ meet.

The proof is now complete.

COROLLARY 5.2. *Let G be a finite graph with set of nodes N. There exists a matching in G with cardinality p if and only if*

$$p(T) \leqslant |T| + (|N| - 2p) \qquad (T \subseteq N),$$

where $p(T)$ is the number of odd components of $G_{N\setminus T}$.

This corollary, derived by Berge [2] as a generalization of a theorem of Tutte (Theorem 5.3), follows from the theorem by taking $A = N$, and $t = 2p$. For, a matching with cardinality p is equivalent to a matching that $2p$ nodes of N meet.

THEOREM 5.3. (Tutte [30, 31]). *For a finite graph G with node set*

N, there exists a perfect matching if and only if

$$p(T) \leqslant |T| \qquad (T \subseteq N). \tag{5.3}$$

This theorem follows from the preceding corollary. For, by taking $T = \emptyset$, we see from (5.3) that G has an even number of nodes. Now take $2p = |N|$.

From the proof of Theorem 5.1, one can obtain some information about the nature of those matchings in a graph which meet a given set of nodes in maximum cardinality. The theorem which we now give extends a result of Edmonds [9] (quoted in [10]) which follows from his algorithm for obtaining maximum cardinality matchings in a graph.

THEOREM 5.4. *Let G be a finite graph with node set N and let $A \subseteq N$. Let t be the maximum cardinality of a set of nodes in A which meets a matching in G. Let M be the collection of subsets of A which meet some matching in G. Let $A^* = \cap\{B : B \in M, |B| = t\}$. Then the node sets $(A_j : j \in J)$ of the connected components of the graph $G_{A \setminus A^*}$ all have odd cardinality. Moreover, if $T \subseteq N \setminus (\cup_{j \in J} A_j)$ is the set of nodes not in $\cup_{j \in J} A_j$ which are joined to at least one node in $\cup_{j \in J} A_j$, then every matching in G which t nodes of A meet contains $\frac{1}{2}(|A_j| - 1)$ edges in the graph G_{A_j} meeting A_j in any prescribed subset of cardinality $|A_j| - 1$ and an edge joining each node of T to a node in $\cup_{j \in J} A_j$.*

This theorem follows from the happenings in the proof of Theorem 5.1. The integer t in the theorem is given by

$$t = \min\{|A| + |T| - p(T; A) : T \subseteq N\}.$$

Choose a $T \subseteq N$ for which the minimum is attained. The nodes of A in the connected components of $G_{N \setminus T}$ which have an even number of nodes or else an odd number of nodes not all of which belong to A are nodes in A^*. Likewise, all nodes of A in T are in A^*. The variability in which subsets of A meet matchings in G comes from the nature of the matchings in the bipartite graph which we constructed in the proof of Theorem 5.1.

Just as we were able to extend results on matchings in finite bipartite graphs to infinite bipartite graphs, provided certain nodes had finite degree, we can extend our results on matchings to infinite graphs. We will again make use of the selection principle of Rado, Theorem 4.1.

THEOREM 5.5. *Let G be a graph with N its set of nodes, and let $A \subseteq N$. Assume that each node in A has finite degree. Then A meets a matching in G if and only if every finite subset of A meets some matching in G.*

Let N_a be the finite set of nodes which are joined by an edge to a $(a \in A)$. Suppose for each $F \subset\subset A$ there is a matching which F meets. Thus there is a choice function θ_F of $(N_a : a \in F)$ such that $\theta_F(a) \in N_a (a \in A)$ where $\theta_F(a_1) \neq \theta_F(a_2)$ $(a_1, a_2 \in F, a_1 \neq a_2)$ and where if $\theta_F(a) = b \in F$ then $\theta_F(b) = a$. Let θ be the choice function of $(N_a : a \in A)$ whose existence is given by Theorem 4.1. Let Δ be the set of edges which join a and $\theta(a)$ $(a \in A)$. We show that Δ is a matching, from which the theorem follows. Let $a_1, a_2 \in A$ with $a_1 \neq a_2$. If $F = \{a_1, a_2\}$, then there is an E with $F \subseteq E \subset\subset A$ such that

$$\theta(a_i) = \theta_E(a_i) \qquad (i = 1, 2).$$

Thus $\theta(a_1) \neq \theta(a_2)$ and if $\theta(a_1) = a_2$, then $\theta(a_2) = a_1$. This is equivalent to Δ being a matching.

If we combine Theorem 5.5 with Corollary 5.1, we obtain the following theorem:

THEOREM 5.6. *Let G be a graph with N its set of nodes, and let $A \subseteq N$. Assume that each node in A has finite degree. Then A meets a matching in G if and only if*

$$p(T; A) \leqslant |T| \qquad (T \subset\subset N). \tag{5.4}$$

The condition (5.4) is easily seen to be necessary. Suppose now (5.4) holds. Let $F \subset\subset A$ and let F^* be the finite set consisting of the nodes in F and all nodes of G which are joined by an edge to some node in F. There is a matching in G_{F^*} that F meets if and only if there is such a matching in G. According to Corollary 5.2 the former is the case if and only if

$$p(G_{F^*} : T; F) \leqslant |T| \qquad (T \subseteq F^*). \tag{5.5}$$

But clearly for each $T \subseteq F^*$,

$$p(G_{F^*} : T; F) \leqslant p(G : T; A).$$

Thus since (5.4) holds, so does (5.5). Hence for all $F \subset\subset A$, there is a matching in G_{F^*} (and hence in G) which F meets. The conclusion now follows from Theorem 5.5.

As a corollary we obtain a theorem of Tutte [32].

COROLLARY 5.7. *Let G be a graph with N its set of nodes, and suppose every node has finite degree. Then there is a perfect matching in G if and only if*

$$p(T) \leqslant |T| \qquad (T \subset\subset N)$$

where $p(T)$ is the number of odd components of $G_{N \setminus T}$.

6. TRANSVERSAL AND MATCHING MATROIDS

In this section we want to explore the concept of a matroid on a set and its relation to the primary subject of this article, transversal theory. We shall develop only enough of the theory of matroids which will be sufficient for our purposes and which hopefully will enable the reader to put the theorems in reasonable perspective. The concept of a matroid abstracts the properties of linear independence of points in a projective or affine space. Because of this it is also referred to as a combinatorial geometry. It has become clear recently that the concept of a matroid or combinatorial geometry is very pervasive in combinatorial theory, and it has had a unifying influence on many combinatorial situations. The pioneering work on matroids was done by Whitney [35] in 1935, while Tutte [33, 34] has vigorously developed its theory proving many profound theorems. A systematic development of the theory can also be found in [8].

Let E be a set. A *matroid* M on E is a non-empty collection of subsets of E which satisfy the following two properties:

$$A \in M,\ A' \subseteq A \qquad \text{imply } A' \in M. \tag{6.1}$$

$$A_1, A_2 \in M, \quad |A_1| + 1 = |A_2| < \infty \quad \text{imply there is an } x \in A_2 \backslash A_1 \text{ such that } A_1 \cup \{x\} \in M. \tag{6.2}$$

Because a matroid M is non-empty, we have in view of (6.1) that $\emptyset \in M$. These properties are familiar to all of us who have studied affine or linear spaces over a field or division ring and the notion of linear independence of points or vectors. Indeed, if V is such a linear space and $E \subseteq V$, then the collection M of all subsets of E which are linearly independent sets of vectors is a matroid on E†. In a vector space an infinite set of vectors is defined to be linearly independent provided every finite subset of these vectors is. We say that the matroid M on E is a *finite-character* matroid if the following additional property is satisfied:

$$\text{For } A \subseteq E,\ A \in M \text{ if } A' \in M \text{ for all } A' \subset\subset A. \tag{6.3}$$

Of course, every matroid on a finite set is a finite-character matroid. If M is a matroid on a set E, the members of M are referred to as the *independent sets*; all other subsets of E are *dependent sets*.

Matroids arise in many ways. We have already mentioned those arising from linear spaces. From set theory we get the following examples: Let E be a set and n a non-negative integer. The

---

†Not all matroids arise this way. Such matroids are called *linear*. The first example of a non-linear matroid is due to Mac Lane [20].

collection $\mathcal{P}(E)$ of all subsets of E and the collection $\mathcal{P}_n(E)$ of all subsets of E with at most n elements are finite-character matroids, while the collection $\mathcal{P}^c(E)$ of all subsets of E which are finite or countably infinite is an example of a matroid which does not have the finite-character property (6.3) if E is an uncountable set.

An example of a matroid from the theory of graphs is the following. Let E be the set of edges of a finite graph G and let M be the collection of subsets of E which do not contain the set of edges of any cycle in G. For these 'graphic' matroids, the independent sets are the edge sets of subgraphs of G which are forests. The edge sets of simple cycles are the minimal dependent sets. If we take the graph in Figure 4, then $E = \{a, b, c, d, e\}$ and M consists of the empty set, all singletons, all doubletons, and all three element subsets of E except for $\{a, b, e\}$ and $\{c, d, e\}$. There are many other realizations of matroids some of which are the subject of this section.

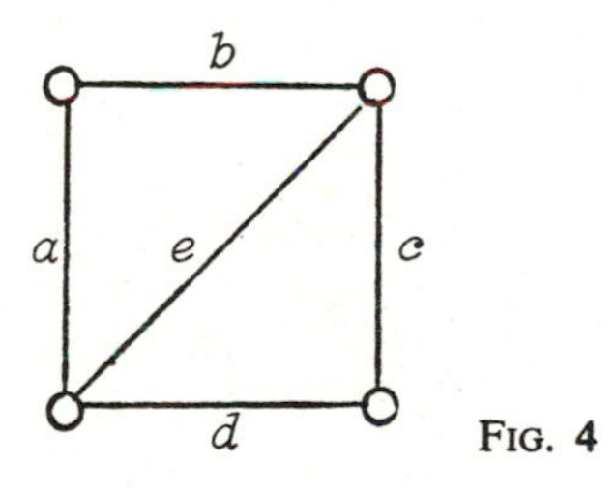

FIG. 4

Let M be a matroid on a set E. A *rank function* ρ is defined on subsets A of E in the following way. If for each positive integer n there is an independent subset of E with at least n elements, we set $\rho(A) = \infty$. Otherwise, the number of elements in independent subsets of A is bounded above, and we set $\rho(A) = m$ if m is the largest number of elements in an independent subset of A. A *basis* of M is any independent subset B of E which is not properly contained in any other independent set. Thus if the members of M are partially ordered by containment, the bases of M (if there are any) are the maximal members of M with respect to the partial order. The matroid $\mathcal{P}^c(E)$ where E is an uncountable set has no bases. Matroids on finite sets, of course, have bases; moreover, every independent set can be enlarged to a basis. The same is true for finite-character matroids, and this can be easily established by use of Zorn's lemma.

LEMMA 6.1. *Let M be a matroid on E and let $A \subseteq E$ with $\rho(A) = m < \infty$. Suppose $B \subseteq A$ with $B \in M$ and $|B| = m$ and suppose $F \subseteq A$ with $F \in M$. Then there exists $X \subseteq B \setminus F$ such that $F \cup X \in M$ and $|F \cup X| = m$. Thus if $\rho(E) < \infty$, any two bases of M have the same number of elements.*

If $|F| = |B|$ we may take $X = \emptyset$. Otherwise $|F| < |B|$ and repeated application of property (6.2) furnishes the desired result.

There are three constructions for matroids that we are going to make use of. The first two of these are rather simple. (1) Let M^i be a matroid on E^i $(i = 1, 2)$ where $E^1 \cap E^2 = \emptyset$. Then $M^1 \oplus M^2 = \{A^1 \cup A^2 : A^1 \in M^1, A^2 \in M^2\}$ is a matroid on $E^1 \cup E^2$, called the *direct sum* of M^1 and M^2. (2) Let M be a matroid on E and let $A \subseteq E$, then $M_A = \{F : F \subseteq A, F \in M\}$ is a matroid on A, called the *restriction* of M to A. If M satisfies the finite character property so does M_A. (3) Let M be a matroid on the finite set E and let $A \subseteq E$. Let B be a basis of $M_{E \setminus A}$ (that is, a maximal independent subset of $E \setminus A$), then $M_{\otimes A} = \{F : F \subseteq A, F \cup B \in M\}$ is a matroid on A, called the *contraction* of M to A. When considering a contraction of a matroid, one would do well to keep in mind the idea of taking the quotient of a vector space by a subspace.

That the above three constructions lead to matroids is easily verified. What is not so obvious is that the matroid $M_{\otimes A}$ does not depend on the choice of the basis B of $M_{E \setminus A}$. This is the content of the following lemma.

LEMMA 6.2. *Let M be a matroid on the finite set E. Let $A \subseteq E$ and let B_1, B_2 be bases of the matroid $M_{E \setminus A}$. Let $F \subseteq A$ and suppose $F \cup B_1 \in M$. Then $F \cup B_2 \in M$.*

Since B_1, B_2 are bases of $M_{E\setminus A}$, for $y \in B_1 \setminus B_2$, $B_2 \cup \{y\}$ and thus any set containing it is a dependent set. Likewise, for $x \in B_2 \setminus B_1$, $B_1 \cup \{x\}$ and any set containing it is a dependent set. Therefore $F \cup B_1$ is a basis of the matroid $M_{F \cup B_1 \cup B_2}$. Suppose $F \cup B_2$ were not an independent set. Then there exists $F' \subsetneq F$ such that $F' \cup B_2$ is a maximal independent subset of $F \cup B_2$. Thus $\{y\} \cup F' \cup B_2 \notin M$ for all $y \in F \setminus F'$. But then $\{y\} \cup F' \cup B_2 \notin M$ for all $y \in (F \cup B_1 \cup B_2) \setminus (F' \cup B_2)$ so that $F' \cup B_2$ is a basis of $M_{F \cup B_1 \cup B_2}$. But since $|F' \cup B_2| < |F \cup B_1|$, this contradicts the conclusion of Lemma 6.1. Hence $F \cup B_2 \in M$.

Let ρ^1, ρ^2, ρ denote the rank functions of the matroids M^1, M^2, M respectively. Then the rank of $F \subseteq E^1 \cup E^2$ with respect to the matroid $M^1 \oplus M^2$ is $\rho^1(F \cap E^1) + \rho^2(F \cap E^2)$. The rank function of M and M_A agree wherever they are both defined (on subsets of A) and we do not distinguish them. If ρ_A denotes the rank function of $M_{\otimes A}$, then for $F \subseteq A$, $\rho_A(F) = \rho(F \cup (E \setminus A)) - \rho(E \setminus A)$.

The first* connection between transversal theory (or, more generally, matching theory in graphs) and matroid theory is brought out in the following theorem of Edmonds and Fulkerson [10].

THEOREM 6.3. *Let G be a graph with N its set of nodes. Let X be a fixed subset of nodes. Let $M(G, X)$ consist of those subsets of X which meet some matching in G. Then $M(G, X)$ is a matroid on X. If every node of X has finite degree, then $M(G, X)$ is a finite-character matroid.*

Any matroid that arises this way is called a *matching matroid*. Before proving the theorem, we wish to state as a special result, the following corollary which is a special case of the theorem.

COROLLARY 6.4. *Let $\mathfrak{A} = (A_i : i \in I)$ be a family of subsets of a set E. Let $M(\mathfrak{A})$ be the collection of partial transversals of $\mathfrak{A}$. Then $M(\mathfrak{A})$ is a matroid on E. If each element of E is a member of only*

---

*We are not speaking historically here.

finitely many sets of the family, then $M(\mathfrak{A})$ is a finite-character matroid.

The corollary follows from Theorem 6.3 by associating with the family the bipartite graph (I, Δ, E) as before and then taking $X = E$. Matroids whose independent sets are the partial transversals of some family of sets are called *transversal matroids*. Every transversal matroid is a matching matroid. The converse, surprisingly perhaps, is also true for finite character matroids [10]. We shall prove it later for matroids on a finite set. Just as the set of partial transversals of the family $\mathfrak{A}$ is a matroid on E, the collection of subsets J of I such that $(A_i : i \in J)$ has a transversal is a matroid on I. This follows by taking $X = I$ in Theorem 6.3 applied to the graph (I, Δ, E). This matroid is just the transversal matroid of the dual family $(A_e^* : e \in E)$ where $A_e^* = \{i \in I : e \in A_i\}$.

In order to prove Theorem 6.3 it is enough to prove that the collection $M = M(G, N)$ of subsets of N which meet a matching is a matroid. This is because $M(G, X) = M(G, N)_X$. To prove that M is a matroid we need only verify property (6.2) of a matroid, (6.1) being obvious. So let A_1, A_2 be finite members of M with $|A_1| < |A_2|$. Let Δ_i be a matching which meets A_i $(i = 1, 2)$, and consider the subgraph G^* of G whose edges are the edges in $\Delta_1 \cup \Delta_2$ and whose nodes are the nodes of G incident with at least one edge in $\Delta_1 \cup \Delta_2$. Then each node of G^* is incident with either one or two edges of G^* so that the connected components of G^* are either cycles or paths joining two distinct nodes, in each case the edges alternating between Δ_1 and Δ_2.

Suppose first there is an edge of Δ_1 which meets a node x of $A_2 \setminus A_1$. Then Δ_1 is a matching which meets $A_1 \cup \{x\}$ and property (2) of a matroid is satisfied. Now suppose that no edge of Δ_1 meets a node of $A_2 \setminus A_1$. Then since $|A_2| > |A_1|$, there is a connected component of G^* which is a path P such that all nodes of A_1 that are nodes of P are nodes of A_2 and, in addition, there is a node x of P which is a node of $A_2 \setminus A_1$. If Δ_i^x is the set of edges of Δ_i which are edges of the path P $(i = 1, 2)$, then $(\Delta_1 \setminus \Delta_1^x) \cup \Delta_2^x$ is a matching in G which meets $A_1 \cup \{x\}$. Thus property (2) of a matroid is satisfied and M is a matroid.

Suppose now $X \subseteq N$ and each node of X has finite degree. We want to show that $M(G, X)$ is then a finite character matroid. So let $A \subseteq X$ and suppose every finite subset of A meets a matching in G. According to Theorem 5.5, A itself meets a matching in G so that $A \in M(G, X)$. This completes the proof of Theorem 6.2.*

Theorem 5.1 furnishes an explicit description of the rank function of the matching matroid that results from a finite graph. Precisely, if G is a finite graph with node set N, $A \subseteq N$, and ρ is the rank function of the matroid $M(G, N)$, then

$$\rho(A) = \min\{|A| + |T| - p(T; A) : T \subseteq N\},$$

where we recall that $p(T; A)$ is the number of odd components of $G_{N \setminus T}$ all of whose nodes belong to A. In the same way Corollary 1.3 gives explicitly the rank function of a transversal matroid. If $\mathfrak{A} = (A_i : i \in I)$ is a finite family of subsets of E, $A \subseteq E$, and ρ is the rank function of the transversal matroid $M(\mathfrak{A})$, then

$$\rho(A) = \min\left\{ \left|\left(\bigcup_{i \in J} A_i \right) \cap A\right| + |I \setminus J| : J \subseteq I \right\}.$$

This follows by applying Corollary 1.3 to the family $(A_i \cap A : i \in I)$.

We now turn to the problem of showing that every finite matching matroid is a transversal matroid. Since a restriction of a matching matroid is also a matching matroid (this follows from the definition of a matching matroid), and since a restriction of a transversal matroid is also a transversal matroid (if $\mathfrak{A} = (A_i : i \in I)$ is a family of subsets of E and $X \subseteq E$, then $M(\mathfrak{A})_X$ is the transversal matroid of $(A_i \cap X : i \in I)$), it is enough to show that if G is a finite graph then $M(G, N)$ is a transversal matroid.

We shall find it useful to introduce one new concept. If G is a connected graph, an edge is an *isthmus* provided the deletion of that edge separates the graph. An isthmus has the property that it must be an edge of every spanning tree of G (a tree whose set of nodes is the set of nodes of G and whose edges are edges of G).

---

*It is suggested that the reader work through the above proof for the special case of a bipartite graph.

The edge sets of spanning trees are the bases of the graphic matroid associated with G. If M is a matroid on a finite set E and $x \in E$, then x is an *isthmus* of M provided x is in every basis of M (equivalently $A \cup \{x\} \in M$ whenever $A \in M$).

LEMMA 6.5. *If M is a matroid on a finite set E and F is a set of isthmuses, then M is a transversal matroid if $M_{E \setminus F}$ is.*

Let $M_{E \setminus F}$ be the transversal matroid of the family $\mathfrak{A} = (A_i : i \in I)$ of subsets of $E \setminus F$. Let K be a set with $|K| = |F|$ and $K \cap I = \emptyset$, and consider the family $\mathfrak{B} = (B_j : j \in K \cup I)$ where $B_j = F$ if $j \in K$ and $B_j = A_j$ if $j \in I$. Since each basis of M is the union of a basis of $M_{E \setminus F}$ and F, M is the transversal matroid of $\mathfrak{B}$.

LEMMA 6.6. *Let M be a matroid on the finite set E and suppose F is a subset of E with the property that every basis of M contains all but at most one element of F. Let M^* be the matroid on $(E \setminus F) \cup \{x\}$ where x is an element not in $E \setminus F$ whose bases are the sets $(B \setminus F) \cup \{x\}$ if B is a basis of M containing F and $B \setminus F$ if B is a basis of M containing all but one element of F. Then if M^* is a transversal matroid, so is M.*

We first show that M^* is really a matroid on $(E \setminus F) \cup \{x\}$, and for this we need only verify property (6.2) of a matroid. Thus let $A_1, A_2 \in M^*$ with $|A_1| + 1 = |A_2|$. We distinguish several cases.

CASE 1. Suppose $x \in A_1 \cap A_2$. Then $A_1' = (A_1 \setminus \{x\}) \cup F$ and $A_2' = (A_2 \setminus \{x\}) \cup F$ are in M. Hence there exists $y \in A_2' \setminus A_1' (= A_2 \setminus A_1)$ such that $A_1' \cup \{y\} \in M$. Thus $A_1 \cup \{y\} \in M^*$.

CASE 2. Suppose $x \in A_1 \setminus A_2$. Then $A_1' = (A_1 \setminus \{x\}) \cup F \in M$, while $A_2' = A_2 \cup (F \setminus \{f\}) \in M$ for some $f \in F$. Thus there exists $y \in A_2' \setminus A_1' (= A_2 \setminus A_1)$ such that $A_1' \cup \{y\} \in M$. Thus $A_1 \cup \{y\} \in M^*$.

CASE 3. Suppose $x \in A_2 \setminus A_1$. Then $A_1' = A_1 \cup (F \setminus \{f\}) \in M$ for some $f \in F$ while $A_2' = (A_2 \backslash \{x\} \cup F \in M$. Hence either $A_1' \cup \{x\} \in M$, in which case $A_1 \cup \{x\} \in M^*$ or $A_1' \cup \{e\} \in M$ for some $e \in A_2 \backslash A_1$, $e \neq x$, in which case $A_1 \cup \{e\} \in M^*$.

CASE 4. Suppose $x \not\in A_1 \cup A_2$. Then A_1, $A_2 \subseteq E \backslash F$ so that A_1, $A_2 \in M$. Thus there exists $e \in A_2 \backslash A_1$ such that $A_1 \cup \{e\} \in M$ with $A_1 \cup \{e\} \subseteq E \backslash F$. But then $A_1 \cup \{e\} \in M^*$.

Thus M^* is a matroid. Suppose now M^* is the transversal matroid of the family $(A_i : i \in I)$ of subsets of $(E \backslash F) \cup \{x\}$. We can assume no element of F is an isthmus because of Lemma 6.5. Let K be a set with $K \cap I = \emptyset$ and $|K| = |F| - 1$. Define a family $\mathfrak{A}^* = (A_i^* : i \in K \cup I)$ by $A_i^* = (A_i \backslash \{x\}) \cup F$ is $x \in A_i$ and $i \in I$, $A_i^* = A_i$ if $x \not\in A_i$ and $i \in I$, and $A_i^* = F$ if $i \in K$. It is now a straightforward matter to verify that M is the transversal matroid of the family $\mathfrak{A}^*$.

THEOREM 6.7 (Edmonds and Fulkerson [10]). *Let G be a finite graph with node set N and let $X \subseteq N$. Then the matching matroid $M(G, X)$ is a transversal matroid on X.*

As already pointed out, it suffices to show $M(G, N)$ is a transversal matroid. To prove this we make use of Theorem 5.4 which describes the structure of the matchings in G of maximum cardinality, say p. According to the theorem if N^* is the set of all nodes which meet every matching of cardinality p, that is the set of isthmuses of $M(G, N)$,† then the node sets $N_j (j \in J)$ of the connected components of $G_{N \backslash N^*}$ all have odd cardinality. Moreover, if $T \subseteq N \backslash N^*$ is the set of nodes which are joined to at least one node in $\cup_{j \in J} N_j$, then every matching in G which has cardinality p contains $\frac{1}{2}(|N_j| - 1)$ edges in the graph G_{N_j} and an edge joining each node of T to a node in $\cup_{j \in J} A_j$. By Lemma 6.5, we need only show $M_{N \backslash N^*}$ is a matroid. But according to the preceding remarks, each basis of $M_{N \backslash N^*}$ (or of M) contains all but

†This matroid has rank $2p$.

at most one node in $N_j (j \in J)$. By repeated application of Lemma 6.6, we find that M^* is a matroid on J where for $K \subseteq J$, K is an independent set of M^* if and only if there is an independent set of $M_{N \setminus N^*}$ which contains N_j for each $j \in K$. Moreover, we know from Lemma 6.6 that $M_{N \setminus N^*}$ is a transversal matroid if M^* is. Let $\mathfrak{A} = (A_t : t \in T)$ be the family of subsets of J such that $j \in A_t$ if and only if there is an edge of G joining t and a node in N_j. Then M^* is the transversal matroid of $\mathfrak{A}$. Hence M^*, thus $M_{N \setminus N^*}$, thus M are all transversal matroids.

The preceding theorem shows that every matching matroid on a finite set is a transversal matroid. More generally, finite-character matching matroids are transversal matroids. It is also well known that transversal matroids (or finite-character transversal matroids) are linear matroids, that is, one can identify the elements with vectors in a vector space so that the independent sets of the matroids correspond exactly to the linearly independent sets of vectors. To see this let M be the transversal matroid of the finite family $\mathfrak{A} = (A_i : i \in I)$ of subsets of the finite set E, and let B be the matrix whereby each $i \in I$ $(e \in I)$ corresponds to a row (column) and a 1 is placed at the (i, e) position if $e \in A_i$ and a 0 is placed otherwise. Let the number of 1's in B be k and let $z_1, \ldots, z_t$ be algebraically independent elements over the rational number field Q. Replace each 1 of A by one of $z_1, z_2, \ldots, z_t$ so that all of them are used. A set of columns of the resulting matrix B^* is a linearly independent set over the field $Q(z_1, \ldots, z_k)$ of rational functions in $z_1, \ldots, z_k$ with coefficients in Q if and only if the corresponding members of E comprise a partial transversal of $\mathfrak{A}$. One can carry this a bit further to realize $M(\mathfrak{A})$ in a linear space over Q (or any sufficiently large field) by letting $q(z_1, \ldots, z_k)$ be the polynomial with integral coefficients which is the product of all non-zero determinants of square submatrices of B^*. There exist rational numbers $r_1, \ldots, r_k$ such that $q(r_1, \ldots, r_k) \neq 0$. If we replace z_i by $r_i (1 \leqslant i \leqslant k)$ in B^*, the columns of the resulting matrix give a realization of $M(\mathfrak{A})$ in a linear space over Q [23].

A good characterization of transversal matroids can be found in [7].

7. TRANSVERSAL THEORY AND MATROIDS

In the preceding section we have seen one way in which the theory of matroids bears on transversal theory–the collection of partial transversals of a family of sets is a matroid. This already sheds some light on some problems in transversal theory. For instance, consider the problem of finding a transversal of a finite family $\mathfrak{A} = (A_i : i \in I)$ of subsets of a set E which contains a prescribed subset M of E. In order for such a transversal to exist, it is first necessary that $\mathfrak{A}$ have a transversal and then it is necessary that M be a partial transversal of $\mathfrak{A}$. But this already suffices, for then the transversals of $\mathfrak{A}$ are the bases of the transversal matroid $M(\mathfrak{A})$ while the partial transversals of $\mathfrak{A}$ are the independent sets of $M(\mathfrak{A})$. If M is a partial transversal of $\mathfrak{A}$, it is an independent set of $M(\mathfrak{A})$, thus can be enlarged to a basis of $M(\mathfrak{A})$, which is then a transversal of $\mathfrak{A}$ containing M. If we now invoke Theorems 1.1 and 1.4, we have proved the following theorem of Hoffman and Kuhn [18]:

THEOREM 7.1. *Let $\mathfrak{A} = (A_i : i \in I)$ be a finite family of subsets of a set E and let $M \subseteq E$. There is a transversal of $\mathfrak{A}$ containing M if and only if the following two conditions are satisfied*:

$$\left|\bigcup_{i\in K} A_i\right| \geqslant |K| \qquad (K \subseteq I). \tag{1}$$

$$\left|\left(\bigcup_{i\in K} A_i\right) \cap M\right| + |I \setminus K| \geqslant |M| \qquad (K \subseteq I). \tag{2}$$

But the theory of matroids bears on transversal theory in another significant way. This was first recognized by Rado [27] who considered the following problem: Suppose $\mathfrak{A} = (A_i : i \in I)$ is a finite family of subsets of a set E and M is a matroid on E. When does $\mathfrak{A}$ have a transversal T which is an independent set of the matroid M? Rado answered this with the following theorem:

THEOREM 7.2. *Let $\mathfrak{A} = (A_i : i \in I)$ be a finite family of subsets of a set E and let M be a matroid on E with rank function ρ. Then $\mathfrak{A}$ has a transversal T with $T \in M$ if and only if*

$$\rho\left(\bigcup_{i \in K} A_i\right) \geqslant |K| \qquad (K \subseteq I).$$

Theorem 1.1 is the special case of Theorem 7.2 obtained by taking $M = \mathcal{P}(E)$, the collection of all subsets of E. The rank function of $\mathcal{P}(E)$ is just the ordinary cardinality function.

Formulated in the language of bipartite graphs, Theorem 1.2 becomes the following: We have a finite bipartite graph $\langle X, \Delta, Y\rangle$ and a matroid M on Y with rank function ρ. We seek a matching Δ' which matches X with a subset of Y which is an independent set in M. Theorem 7.2 then can be shown to be equivalent to: The matching Δ' exists if and only if

$$|Z \cap X| + |\rho(Z \cap Y)| \geqslant |X|$$

for all sets of nodes Z which separate X and Y. In this form the theorem of Rado is unsymmetrical, for while we are given a matroid on Y and want our matching Δ' to take this into account, we make no such assumption for X†. This non-symmetrical character of Rado's theorem is eliminated by the following theorem:

THEOREM 7.3. *Let $\langle X, \Delta, Y\rangle$ be a finite bipartite graph. Let M^1 be a matroid on X with rank function ρ^1 and M^2 a matroid on Y with rank function ρ^2. Let t be a non-negative integer. Then there exists a matching Δ^* of cardinality t which matches an independent set of M^1 with an independent set of M^2 if and only if*

$$\rho^1(Z \cap X) + \rho^2(Z \cap Y) \geqslant t \tag{7.1}$$

for all separating sets Z. Thus the maximum cardinality of a

†Of course, we could say that we have the matroid $\mathcal{P}(X)$ on X.

matching Δ^* *with these properties is*

$$\min\{\rho^1(Z \cap X) + \rho^2(Z \cap Y) : Z \text{ a separating set}\}.$$

Condition (7.1) is equivalent to

$$\rho^1(X \backslash A) + \rho^2(A^\Delta) \geqslant t \qquad (A \subseteq X),$$

where for $A \subseteq X$, A^Δ is the set of all nodes in Y which are joined by an edge to at least one node in A. The set $(X \backslash A) \cup A^\Delta$ is always a separating set, so that what we are saying is that only this kind of separating set need be considered.

We have purposely not proved the three preceding theorems, for we want to prove a more general theorem of which they are special cases. All three of the theorems do admit direct proofs.

Our generalization is motivated by the classical theorem of Menger [22]. Let G be a finite directed graph and let X, Y be disjoint subsets of the nodes of G. Menger's theorem† asserts that *the maximum cardinality of a set of pairwise node disjoint paths from* X *to* Y *equals the minimum cardinality of a set of nodes which separates* X *from* Y. In the directed graph in Figure 5, $\{z_1, z_2\}$ is a set of two nodes which separates X from Y, while one easily finds two node disjoint paths from X to Y. Theorem 2.1 is a special case of Menger's theorem where to be precise we should think of the edges of a bipartite graph $\langle X, \Delta, Y\rangle$ as all being directed from X to Y.

We want to impose matroid structures on X and Y and require that the set of initial nodes of the paths and the set of terminal nodes be independent sets in the respective matroids. We now make this precise.

A *finite directed graph* has a finite set N of nodes along with a set of edges which are ordered pairs (x, y) of distinct nodes. The edge (x, y) is regarded as having *initial node* x and *terminal node* y. By a *path* θ in G we mean a linearly ordered sequence

†Menger formulated his theorem for undirected graphs. The generalization to directed graphs is proved by Ford and Fulkerson [12] from their powerful max-flow-min cut network theorem.

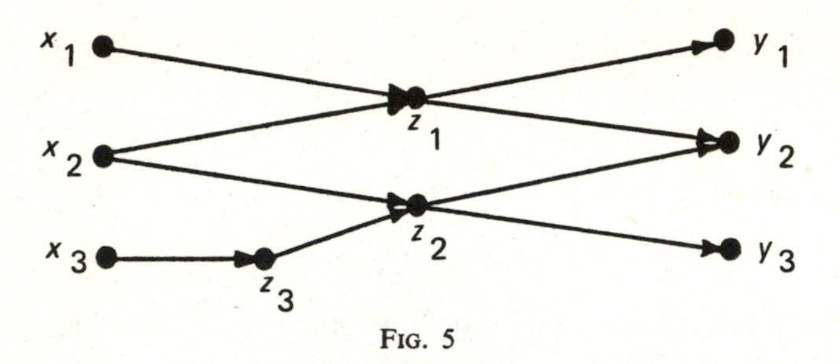

FIG. 5

$(x_1, x_2, \ldots, x_n)$ of $n \geqslant 2$ distinct nodes with (x_i, x_{i+1}) an edge of G $(1 \leqslant i \leqslant n-1)$. The *initial node* of θ, In θ, is x_1; the *terminal node* of θ, Ter θ, is x_n. The *set of nodes* of θ is Nod $\theta = \{x_1, x_2, \ldots, x_n\}$. If Θ is a collection of paths, In $\Theta = \{\text{In } \theta : \theta \in \Theta\}$ while Ter $\Theta = \{\text{Ter } \theta : \theta \in \Theta\}$. The paths in Θ are *pairwise node disjoint* provided $(\text{Nod } \theta_1) \cap (\text{Nod } \theta_2) = \emptyset$ for $\theta_1, \theta_2 \in \Theta$, $\theta_1 \neq \theta_2$. If $\theta = (x_1, \ldots, x_n)$ and $\pi = (x_n, \ldots, x_m)$ are paths with $(\text{Nod } \theta) \cap (\text{Nod } \pi) = \{x_n\}$, then $\theta * \pi$ is the path $(x_1, \ldots, x_n, \ldots, x_m)$. Suppose now X and Y are disjoint subsets of nodes. A *path θ from X to Y* is a path with In $\theta \in X$, Ter $\theta \in Y$ with no other node of θ in $X \cup Y$. A set Z of nodes *separates X from Y* provided $(\text{Nod } \theta) \cap Z \neq \emptyset$ for each path θ from X to Y.

The following theorem [4] generalizes Menger's theorem:

THEOREM 7.4. *Let G be a finite directed graph with X and Y disjoint subsets of nodes. Let M^1 be a matroid on X with rank function ρ^1 and M^2 a matroid on Y with rank function ρ^2. Then the maximum cardinality of a set Θ of pairwise node disjoint paths with* In $\Theta \in M^1$ *and* Ter $\Theta \in M^2$ *equals*

$$\min\{\mu(Z) : Z \text{ separates } X \text{ from } Y\},$$

where $\mu(Z) = \rho^1(Z \cap X) + \rho^2(Z \cap Y) + |Z \backslash (X \cup Y)|$.

We call $\mu(Z)$ the *index* of the separating set, relative to M^1 and M^2. Note that if $M^1 = \mathcal{P}(X)$, $M^2 = \mathcal{P}(Y)$, the theorem reduces to Menger's theorem since then $\mu(Z) = |Z|$. We now prove Theorem 7.4.

Let Θ be a collection of pairwise node disjoint paths with $\text{In}\,\Theta \in M^1$ and $\text{Ter}\,\Theta \in M^2$, and let Z be a set of nodes which separates X from Y. Set $Z_1 = Z \cap X$, $Z_2 = Z \cap Y$, $Z_0 = Z \setminus (Z_1 \cup Z_2)$. Let Θ_1 (respectively, Θ_2) consist of those paths $\theta \in \Theta$ with $\text{In}\,\theta \in Z_1$ (respectively, $\text{Ter}\,\theta \in Z_2$), and let $\Theta_0 = \Theta \setminus (\Theta_1 \cup \Theta_2)$. Since $\text{In}\,\Theta \in M^1$, $\text{In}\,\Theta_1 \in M^1$ and thus $|\Theta_1| \leqslant \rho^1(Z_1)$. Similarly we find $|\Theta_2| \leqslant \rho^2(Z_2)$. All paths in Θ_0 must contain a node of Z_0, and since paths in Θ and thus Θ_0 are pairwise node disjoint, $|\Theta_0| \leqslant |Z_0|$. Hence

$$|\Theta| \leqslant |\Theta_1| + |\Theta_2| + |\Theta_0| \leqslant \rho^1(Z_1) + \rho^2(Z_2) + |Z_0| = \mu(Z).$$

To complete the proof, we let k be the minimum index of a set Z of nodes which separates X from Y and show by induction on the number of edges of G that there is a set Θ of k pairwise node disjoint paths with $\text{In}\,\Theta \in M^1$ and $\text{Ter}\,\Theta \in M^2$. If G has only one edge this is readily verified. Otherwise we distinguish two cases. We may assume that each node of X is the initial node of at least one edge and each node of Y is the terminal node of at least one edge. We may further assume that there is no edge with both initial and terminal node in X (or in Y), and that $\{x\} \in M^1$, $\{y\} \in M^2$ for each $x \in X, y \in Y$.

CASE 1. Every set Z of nodes which separates X from Y and has index k satisfies $Z \subseteq X$ or $Z \subseteq Y$. In this case, let (x, z) be an edge of G with $x \in X$ (and thus $z \not\in X$) and let G' be the directed graph obtained from G by the removal of the edge (x, z). Consider the graph G', the sets X and Y of nodes, and the matroids M^1 on X and M^2 on Y. If all sets W of nodes which separate X from Y in G' have index at least k, the conclusion of the theorem follows by induction. Otherwise there is a set W of nodes which separates X from Y in G' which has index less than k. But then $W \cup \{x\}$ is a set which separates X from Y in G and clearly $\mu(W \cup \{x\}) \leqslant k$; we must have equality and since $x \in X$ but $x \not\in Y$, $W \cup \{x\} \subseteq X$. Likewise, $W \cup \{z\}$ separates X from Y in G and has index k. Since $z \not\in X$, $W \cup \{z\} \subseteq Y$. But $X \cap Y = \emptyset$ so that $W = \emptyset$, $k = 1$, $\{x\} \in M^1$ and $\{z\} \in M^2$. The path (x, z) satisfies the conclusion of the theorem.

CASE 2. There is a set Z of nodes separating X from Y in G which has index k such that $Z \nsubseteq X$, $Z \nsubseteq Y$. We then put $Z_1 = Z \cap X$, $Z_2 = Z \cap Y$, $Z_0 = Z \setminus (Z_1 \cup Z_2)$, $X_1 = X \setminus Z_1$, and $Y_2 = Y \setminus Z_2$. Then $Y_2 \neq \emptyset$; for if $Y_2 = \emptyset$, then $Y \subseteq Z$ and Y is a separating set with $\mu(Y) < \mu(Z) = k$, which is a contradiction. Likewise, $X_1 \neq \emptyset$. We let G^1 be the graph consisting of the nodes and edges of paths in G from X_1 to $Z_0 \cup Z_2$. The graph G^1 has fewer edges than G, since no edge with a terminal node in Y_2 can be an edge of G^1. Consider then the graph G^1, the disjoint sets of nodes X_1 and $Z_0 \cup Z_2$ and the matroid $(M^1)_{\otimes X_1}$ on X_1 and the matroid $\mathcal{P}(Z_0) \oplus (M^2)_{Z_2}$† on $Z_0 \cup Z_2$. Let the index function for these circumstances be denoted by μ^1, and suppose there were a set of nodes A of G^1 which separates X_1 from $Z_0 \cup Z_2$ in G^1 with $\mu^1(A) < k - \rho^1(Z_1)$. The set $A \cup Z_1$ is then a set of nodes of G which separates X from Y in G. For, let θ be a path in G from X to Y. If $\operatorname{In} \theta \in Z_1$, θ contains a node of $A \cup Z_1$. Otherwise $\operatorname{In} \theta \in X_1$. If $\operatorname{Ter} \theta \in Z_2$, then θ is a path in G^1 from X to $Z_0 \cup Z_2$ and thus θ contains a node of A. If $\operatorname{Ter} \theta \notin Z_2$, then $(\operatorname{Nod} \theta) \cap Z_0 \neq \emptyset$ and an initial subsequence of θ is a path in G^1 from X_1 to $Z_0 \cup Z_2$ so that θ contains a node of A. Thus $A \cup Z_1$ separates X from Y in G. We compute that

$$\begin{aligned}\mu(A \cup Z_1) &= \rho^1((A \cap X_1) \cup Z_1) + \rho^2(A \cap Y) + |A \setminus (X \cup Y)| \\ &= \rho^1(Z_1) + \rho^1_{X_1}(A \cap X_1) + \rho^2(A \cap X) + |A \setminus (X \cup Y)| \\ &= \rho^1(Z_1) + \mu^1(A) < k.\end{aligned}$$

This is a contradiction. Hence $\mu^1(A) \geqslant k - \rho^1(Z_1)$ for all sets of nodes A of G^1 which separate X_1 from $Z_0 \cup Z_2$ in G^1. By induction, we conclude there is a set Θ_1 of $k - \rho^1(Z_1) = \rho^2(Z_2) + |Z_0|$ pairwise node disjoint paths in G^1 from X_1 to $Z_0 \cup Z_2$ with $\operatorname{In} \Theta_1 \in (M^1)_{\otimes X_1}$ and $\operatorname{Ter} \Theta_1 \in \mathcal{P}(Z_0) \oplus (M^2)_{Z_2}$. From cardinality considerations we conclude that $\operatorname{Ter} \Theta_1 = Z_0 \cup B_2$ where B_2 is a basis of $(M^2)_{Z_2}$.

---

†Thus $A \subseteq Z_0 \cup Z_2$ is independent in $\mathcal{P}(Z_0) \oplus (M^2)_{Z_2}$ if $(A \cap Z_2) \in M^2$.

Let G^2 be the graph consisting of the nodes and edges of paths in G from $Z_0 \cup Z_1$ to Y_2. An analogous argument produces a set Θ_2 of $k - \rho^2(Z_2) = \rho^1(Z_1) + |Z_0|$ pairwise node disjoint paths in G^2 from $Z_1 \cup Z_0$ to Y_2 with $\text{In}\,\Theta_2 \in (M^1)_{Z_1} \oplus \mathcal{P}(Z_0)$ and $\text{Ter}\,\Theta_2 \in (M^2)_{\otimes Y_2}$. From cardinality considerations again, $\text{In}\,\Theta_2 = B_1 \cup Z_0$ where B_1 is a basis of $(M^1)_{Z_1}$. The paths in Θ_1 and Θ_2 can have only the nodes in Z_0 in common, for otherwise the separating property of Z_0 is violated. Let Θ be the set of paths which is the union of the following three sets:

(i) $\{\theta_1 : \theta_1 \in \Theta_1, \text{Ter}\,\theta_1 \in Z_2\}$
(ii) $\{\theta_2 : \theta_2 \in \Theta_2, \text{In}\,\theta_2 \in Z_1\}$
(iii) $\{\theta_1 * \theta_2 : \theta_1 \in \Theta_1, \theta_2 \in \Theta_2, \text{Ter}\,\theta_1 = \text{In}\,\theta_2 \in Z_0\}$.

Then Θ is a collection of k pairwise node disjoint paths with $\text{In}\,\Theta = B_1 \cup \text{In}\,\Theta_1 \in M^1$ and $\text{Ter}\,\Theta = B_2 \cup \text{Ter}\,\Theta_2 \in M^2$. The theorem is proved.

If it is allowed that $X \cap Y \neq \emptyset$ and that the paths from X to Y can contain nodes of X and Y other than the initial and terminal nodes respectively, then the notion of a separating set of nodes needs to be replaced by that of a separating triple. A triple of nodes (Z_1, Z_0, Z_2) where $Z_1 \subseteq X$, $Z_2 \subseteq Y$ is a *separating triple* (for X and Y) provided every path θ from X to Y satisfies $\text{In}\,\theta \in Z_1$, $\text{Ter}\,\theta \in Z_2$ or $(\text{Nod}\,\theta) \cap Z_0 \neq \emptyset$. If the index of a separating triple (Z_1, Z_0, Z_2) is defined to be $\rho^1(Z_1) + \rho^2(Z_2) + |Z_0|$, then the preceding theorem remains valid provided degenerate paths, paths consisting of only one node, are permitted. To see that these changes are needed, consider the graph G which has three nodes x_1, x_2, y and edges (x_1, x_2), (x_2, y). Let the matroid M^1 on $X = \{x_1, x_2\}$ have as its independent sets only $\emptyset$ and $\{x_1\}$, and let the matroid M^2 on $Y = \{y\}$ have as its independent sets $\emptyset$ and $\{y\}$. The set $\{x_2\}$ is then a separating *set* with index 0; on the other hand the path $\theta = (x_1, x_2, y)$ satisfies $\{\text{In}\,\theta\} \in M^1$, $\{\text{Ter}\,\theta\} \in M^2$. Examples of separating *triples* of index 1 are $(\{x_1\}, \emptyset, \emptyset)$, $(\emptyset, \{x_2\}, \emptyset)$, and $(\{x_2\}, \{x_2\}, \emptyset)$.

As has already been remarked, Theorems 7.2 and 7.3 are special cases of Theorem 7.4. The curious reader will probably want to take the proof of Theorem 7.4 and see how it simplifies for the special situations of Theorems 7.2 and 7.3.

We now show how Theorem 7.2 can be applied to obtain some results in the transversal theory of two families of sets. Let $\mathfrak{A} = (A_i : i \in I)$ and $\mathfrak{B} = (B_j : j \in J)$ be two families of subsets of E. A set $T \subseteq E$ which is a transversal of both $\mathfrak{A}$ and $\mathfrak{B}$ is called a *common transversal* of $\mathfrak{A}$ and $\mathfrak{B}$. The following theorem of Ford and Fulkerson [11] gives a criterion for the existence of a common transversal of two finite families of sets. The method of proof is due to Mirsky and Perfect [24].

THEOREM 7.5. *Let* $\mathfrak{A} = (A_i : 1 \leqslant i \leqslant n)$ *and* $\mathfrak{B} = (B_j : 1 \leqslant j \leqslant n)$ *be two families of subsets of E. Then* $\mathfrak{A}$ *and* $\mathfrak{B}$ *have a common transversal if and only if*

$$\left|\left(\bigcup_{i \in I} A_i\right) \cap \left(\bigcup_{j \in J} B_j\right)\right| \geqslant |I| + |J| - n \qquad (I, J \subseteq \{1, \ldots, n\}).$$

To prove this theorem, we consider the transversal matroid $M(\mathfrak{B})$. Then $\mathfrak{A}$ and $\mathfrak{B}$ have a common transversal if and only if $\mathfrak{A}$ has a transversal T with $T \in M(\mathfrak{B})$. By Theorem 7.2 this is the case if and only if

$$\rho\left(\bigcup_{i \in I} A_i\right) \geqslant |I| \qquad (I \subseteq \{1, \ldots, n\}) \tag{7.2}$$

where ρ is the rank function of $M(\mathfrak{B})$. But using Corollary 1.3 we find that for $X \subseteq E$,

$$\rho(X) = \min\left\{\left|\left(\bigcup_{j \in J} B_j\right) \cap X\right| + n - |J| : J \subseteq \{1, \ldots, n\}\right\}.$$

Thus (7.2) is equivalent to

$$\left|\left(\bigcup_{j \in J} B_j\right) \cap \left(\bigcup_{i \in I} A_i\right)\right| + n - |J| \geqslant |I| \qquad (J, I \subseteq \{1, \ldots, n\}),$$

and the theorem is proved.

There are many known refinements and generalizations of Theorem 7.4 for both finite and infinite families:

(1) The maximum cardinality of a set which is a partial transversal of two families $\mathfrak{A} = (A_i : 1 \leqslant i \leqslant m)$, $\mathfrak{B} = (B_j : 1 \leqslant j \leqslant n)$ is given by

$$\min\left\{ \left| \left(\bigcup_{i \in I} A_i \right) \cap \left(\bigcup_{j \in J} B_j \right) \right| + n - |I| + m - |J| : \right.$$

$$\left. I \subseteq \{1, \ldots, m\}, J \subseteq \{1, \ldots, n\} \right\}.$$

(2) The families $(A_i : 1 \leqslant i \leqslant n)$, $(B_j : 1 \leqslant j \leqslant n)$ have a common transversal which contains a prescribed set M if and only if

$$\left| \left(\bigcup_{i \in I} A_i \right) \cap \left(\bigcup_{j \in J} B_j \right) \right| + \left| \left(\left(\bigcup_{i \in I} A_i \right) \cup \left(\bigcup_{j \in J} B_j \right) \right) \cap M \right|$$

$$\geqslant |I| + |J| + |M| - n \qquad (I, J \subseteq \{1, \ldots, n\}).$$

(3) If $\mathfrak{A} = (A_i : i \in I)$ and $\mathfrak{B} = (B_j : j \in J)$ are two families of finite subsets of a set E such that each element of E belongs to only finitely many members of $\mathfrak{A}$ and of $\mathfrak{B}$, then $\mathfrak{A}$ and $\mathfrak{B}$ have a common transversal if and only if

$$\left| \left(\bigcup_{i \in K} A_i \right) \cap \left(\bigcup_{j \in J \setminus L} B_j \right) \right| \geqslant |K| - |L| \qquad (K \subset\subset I, L \subset\subset J) \tag{7.3}$$

$$\left| \left(\bigcup_{i \in I \setminus K} A_i \right) \cap \left(\bigcup_{j \in L} B_j \right) \right| \geqslant |L| - |K| \qquad (K \subset\subset I, L \subset\subset J). \tag{7.4}$$

Indeed (7.3) is equivalent to $\mathfrak{A}$ having a transversal T_1 which is a partial transversal of $\mathfrak{B}$, while (7.4) is equivalent to $\mathfrak{B}$ having a transversal T_2 which is a partial transversal of $\mathfrak{A}$. From T_1 and T_2 the desired common transversal can be constructed.

Even further generalizations are possible. For a very general theorem about common transversals, see [3]. Of course Rado's theorem, Theorem 7.2, admits a generalization to infinite families of finite sets, provided the matroid has finite character. The reader who is interested in exploring the subject of transversal theory is encouraged to consult the book *Transversal Theory* by Mirsky [23] and the many references given therein.

But what of common transversals of three finite families of sets, each with the same number of members. While the jump from one family to two families was accomplished without an excessive amount of difficulty, the jump from two families to three families is another matter. At present there are no known conditions which are both necessary and sufficient for three families to have a common transversal. Since the common partial transversals of two families of sets is rarely a matroid, one cannot use the method of proof of Theorem 7.5. If one can pinpoint one reason why the problem of a common transversal of three families of sets is so much more difficult, it is this lack of a matroid structure. This problem of finding a criterion for three finite families of sets to have a common transversal is probably the most outstanding problem in finite transversal theory today.

REFERENCES

1. Banach, S., "Un théorème sur les transformations biunivoques," *Fund. Math.*, **6** (1924), 236–239.

2. Berge, C., *The Theory of Graphs*, London: Methuen, and New York: Wiley, 1962.

3. Brualdi, R. A., "A general theorem concerning common transversals," *Combinatorial Mathematics and Its Applications*, Edited by D. J. A. Welsh. London and New York: Academic Press, 1971, 39–60.

4. ——, "Menger's theorem and matroids," *J. London Math. Soc.*, **2**, 4 (1971), 46–50.

5. ——, "Matchings in arbitrary graphs," *Proc. Cambridge Philos. Soc.*, **69** (1971), 401–407.

6. Brualdi, R. A., and E. B. Scrimger, "Exchange systems, matchings and transversals," *J. Combinatorial Theory*, **5** (1968), 244–257.

7. Brualdi, R. A., and G. W. Dinolt, "Characterizations of transversal matroids and their presentations," *ibid.*, 12 (1972), 268–286.

8. Crapo, H. H., and G. C. Rota, *Combinatorial Geometries* (preliminary edition), Cambridge, Mass: M.I.T. Press, 1970.

9. Edmonds, J., "Paths, trees, and flowers," *Canad J. Math.* **17** (1965), 449–467.

10. Edmonds, J., and D. R. Fulkerson, "Transversals and matroid partition," *J. Res. Nat. Bur. Standards*, Sect. B., **69** (1965), 147–153.

11. Ford, L. R. Jr., and D. R. Fulkerson, "Network flows and systems of representatives," *Canad. J. Math.*, **10** (1958), 78–85.

12. ——, *Flows in Networks*, Princeton: Princeton University Press, 1962.

13. Gallai, T., "Über extreme punkt-und kantenmengen," *Ann. Univ. Sci. Budapest, Eötvös Sect. Math.*, **2** (1959), 133–138.

14. Hall, P., "On representatives of subsets," *J. London Math. Soc.*, **10** (1935), 26–30.

15. Hall, M., Jr., "Distinct representatives of subsets," *Bull. Amer. Math. Soc.*, **54** (1948), 922–926.

16. Halmos, P. R., and H. E. Vaughan, "The marriage problem," *Amer. J. Math.*, **72** (1950), 214–215.

17. Harper, L. H., and G. C. Rota, "Matching theory, an introduction," *Advances in Probability*, **1**. Edited by P. Ney, New York: Marcel Dekker, 1971, 171–215.

18. Hoffman, A. J., and H. W. Kuhn, "On systems of distinct representatives," *Linear Inequalities and Related Systems, Ann. of Math. Study* **38**, Princeton: Princeton University Press, 1956, 199–206.

19. König, D., *Theorie der endlichen und unendlichen graphen*, Leipzig: 1936.

20. Mac Lane, S., "Some interpretations of abstract linear dependence in terms of projective geometry," *Amer. J. Math.*, **58** (1936), 236–240.

21. Mendelsohn, N. S., and A. L. Dulmage, "Some generalizations of the problem of distinct representatives," *Canad. J. Math.*; **10** (1958), 230–241.

22. Menger, K., "Zur allgemeinen Kurventheorie," *Fund. Math.*, **10** (1927), 96–115.

23. Mirsky, L., *Transversal Theory*, New York and London: Academic Press, 1971.

24. Mirsky, L., and H. Perfect, "Applications of the notion of independence to problems of combinatorial analysis," *J. Combinatorial Theory*, **2** (1967), 327–357.

25. Ore, O., "Graphs and matching theorems," *Duke Math J.*, **22** (1955), 625–639.

26. Perfect, H., and J. S. Pym, "An extension of Banach's mapping theorem with applications to problems concerning common representatives," *Proc. Cambridge Philos. Soc.*, **62** (1966), 187–192.

27. Rado, R., "A theorem on independence relations," *Quart. J. Math. Oxford Ser.*, **13** (1942), 83–89.

28. ——, "Axiomatic treatment of rank in infinite sets," *Canad. J. Math.*, **1** (1949), 337–343.

29. ——, "A selection lemma," *J. Combinatorial Theory*, **10** (1971), 176–177.

30. Tutte, W. T., "The factorization of linear graphs," *J. London Math. Soc.*, **22** (1947), 107–111.

31. ——, "The factors of graphs," *Canad. J. Math.*, **4** (1952), 314–328.

32. ——, "The factorization of locally finite graphs," *ibid.*, **2** (1950), 44–49.

33. ——, "Lectures on matroids," *J. Res. Nat. Bur. Standards, Sect. B.*, **69** (1965), 1–47.

34. ——, *Introduction to the theory of matroids*, New York: American Elsevier, 1971.

35. Whitney, H., "On the abstract properties of linear dependence," *Amer. J. Math.*, **57** (1935), 509–533.

ON THE SHORTEST ROUTE THROUGH A NETWORK

George B. Dantzig

The purpose of this paper is to give an efficient procedure for obtaining the shortest route from a given origin to all other nodes in a network or to a particular destination point. The method can be interpreted as a slight refinement of those reported by Bellman, Moore, Ford, and the author in [1], [2], [3], [4], and those proposed by Gale and Fulkerson in informal conversations. It is similar to Moore's method of fanning out from the origin. However, its special feature (which is believed to be new) is that the fanning out is done one point at a time and the distance assigned is final.

It is assumed (a) that one can write down without effort for each node the arcs leading to other nodes in increasing order of length and (b) that it is no effort to ignore an arc of the list if it leads to a node that has been reached earlier. It will be shown that no more than $n(n-1)/2$ comparisons are needed in an n node network to determine the shortest routes from a given origin to all other nodes and less than half this number for a shortest route to a particular node. The basic idea is as follows:

Suppose at some stage k in the computing process the shortest

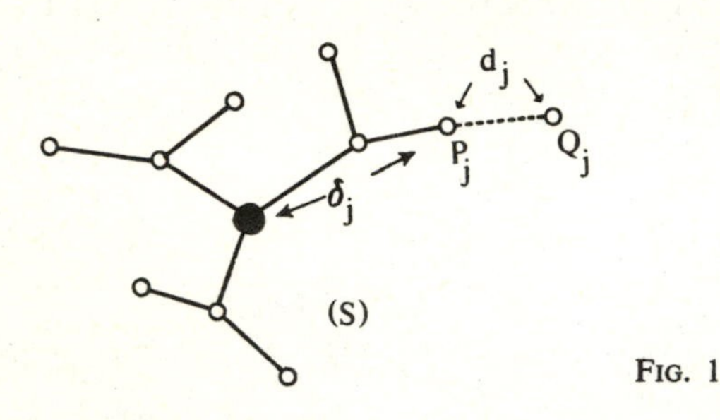

FIG. 1

distances to k of the nodes from some origin are known as well as the paths. Call the set of these points S.

(1) Let P_j be one of the nodes in S,

(2) let δ_j be its least distance to the origin ●,

(3) let Q_j be the nearest node to P_j not in S,

(4) let d_j be the distance from P_j to Q_j. Choose as the $k + 1$ point, Q_s, where s satisfies

$$\delta_s + d_s = \operatorname{Min}(\delta_j + d_j) \qquad j = 1, 2, \ldots, k.$$

The minimum distance of Q_s to the origin is $\delta_s + d_s$ and the best path to the origin is via P_s. The reason is obvious, for if the best path from Q_s were via some other j in S or via several other points not in S and then via some other j in S, then the distance is at least $\delta_j + d_j \geqslant \delta_s + d_s$. In case of ties for minimum, several such nodes Q_s could be determined at the same time and the process made more efficient.

It will be noted that the minimum requires only k comparisons for a decision as to the $(k + 1)$ st point. Hence in an n node network no more than

$$1 + 2 + \cdots + (n - 1) = n(n - 1)/2$$

comparisons are needed.

In practice, the number of comparisons can be considerably less than this bound because after several stages one or more of the nodes in S have only arcs leading to points in S (in the 8 node example below only a total of 16 comparisons was needed instead of 28 comparisons).

If the problem is to determine only the shortest path from a given origin to a given terminal, the number of comparisons may often be reduced by fanning out from both the origin and the terminal simultaneously—adding alternatively one point at a time to sets S about the origin and S' about the terminal.

Once the shortest distance from a node to origin is evaluated, the node is conceptually connected directly to the origin by a hypothetical arc with the specified shortest distance and disconnected from all arcs leading to other nodes evaluated earlier. Nodes whose shortest distance to the terminal which have been determined are similarly treated. Once the origin is reached by either fanning system, the process terminates.

Example.

Distances on links of the network are as indicated in Figure 2. Arrange the nodes in ascending order by distances to a given node:

(O)	(A)	(B)	(C)
OA - 1	AB - 3	BC - 2	CB - 2
OB - 2	AC - 3	BA - 3	CD - 2
	AD - 3	BG - 4	CA - 3
			CG - 3
			CE - 3

(D)	(E)	(F)	(G)
DC - 2	EF - 1	FE - 1	GF - 1
DA - 3	EC - 3	FG - 1	GC - 3
DE - 3	ED - 3		GB - 4

Step 1. *Choose path OA*; place its distance 1 above A column, delete all arcs *into A*.

Step 2. Compare OB-2 and AB-(3 + 1) and *choose path OB*: place its distance 2 above B column, delete all arcs *into B*.

Step 3. Compare AC-(3 + 1), AD-(3 + 1), BC-(2 + 2), and be-

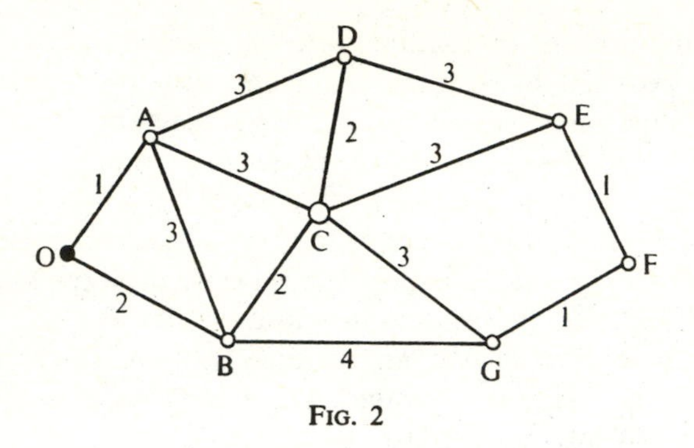

FIG. 2

cause of ties *choose paths AC (or BC) and AD*; place distance 4 above C and D columns, delete all arcs *into C and D*.

Step 4. Compare BG-(4 + 2), CG-(3 + 4), DE-(3 + 4) and *choose path BG*, place its distance 6 above G column, delete all arcs *into G*.

Step 5. Compare CE-(3 + 4), DE-(3 + 4), GF-(1 + 6) and *choose path (CE or DE) and GF*; place distance 7 above E and F columns, delete all arcs *into E and F*.

Because of ties, many of the steps were carried on simultaneously.

The shortest paths from the origin to other nodes are along paths OA, OB, AC, AD, BG, CE, GF, with alternative BC for AC and DE for CE.

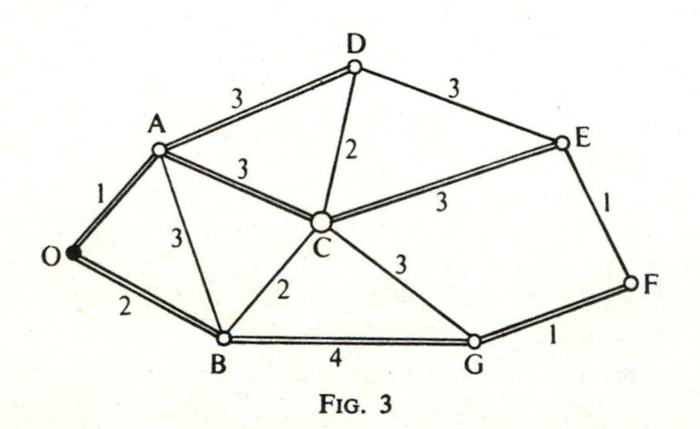

FIG. 3

REFERENCES

1. Bellman, R., "On a routing problem," *Quart. Appl. Math.*, **16** (1958), 87–90.
2. Dantzig, G. B., "Discrete variable extremum problems," *JORSA*, **5** (1957), 266–277.
3. Moore, E. F., "The shortest path through a maze," unpublished mimeographed report, 16 pages.
4. Ford, L. R., Jr., "Network flow theory," *RAND Paper P*-**923** (1956).

ELECTRICAL NETWORK MODELS[1]

R. J. Duffin

INTRODUCTION

The history of science shows that the development of mathematics has been accelerated by the use of models. Thus geometric diagrams have served as models for algebraic relations. Gambling has served as a model for probability theory. Gravitation has served as a model for harmonic functions. Such models have accelerated mathematical development for three main reasons: (i) Attention is focused on significant problems. (ii) Models aid the intuition in perceiving complex relations. (iii) New concepts are suggested.

Since the days of Ohm and Kirchhoff, the study of electrical networks has stimulated developments in practically every branch of mathematics. For example, network models have contributed to

[1]Prepared under Research Grant DA-AROD-31-124-71-G17, Army Research Office, Durham, North Carolina. This is an extension of the paper, "Network models," *SIAM-AMS Proc.* **3** (1971), 65–91.

topology, nonlinear differential equations, function theory, boolean algebra, and information theory.

Networks are still an abundant source of mathematical problems. This paper describes several such problems which have interested me. These problems are varied but will involve the steady flow of electrical current through a network of resistors obeying Ohm's law. This is the classical Kirchhoff network and of main concern here will be the joint resistance of such a network.

1. THE FOUR COLOR PROBLEM

One of the first network models which attracted my interest was an electrical correspondence of map coloration. This can be illustrated by the simple map shown in Figure 1 consisting of a triangular region surrounded by three quadrilateral regions. This map is colored by four "colors," I, II, III, and IV. This coloration has been chosen so that the color IV does not appear on the boundary regions.

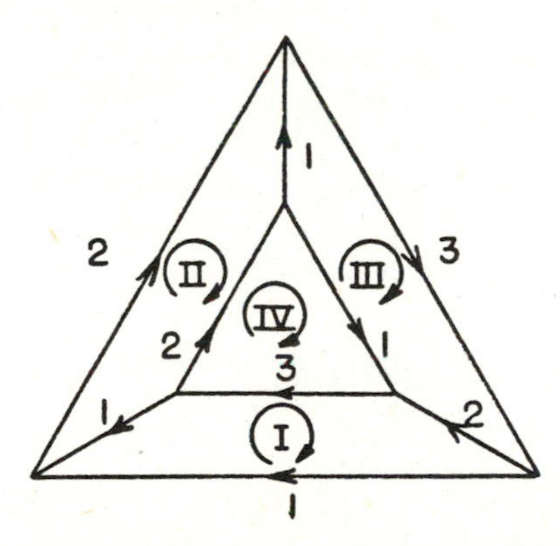

FIG. 1. A map coloration.

If a region has the color number X, let a cyclic current of X amperes flow clockwise around the boundary of the region. Thus IV amperes flows around the boundary of the triangle. Now superimpose all of these cyclic flows. The strength of the resultant flow in the boundary edges is given in arabic numerals, and the direction is indicated by arrows. Thus one edge of the triangle separates regions colored IV and I, so the strength of the current through this edge is IV − I = 3 amperes. Clearly, then, the edge currents have strengths 1, 2, or 3. Moreover, the total current flow satisfies Kirchhoff's current conservation law. This comes about because the total current flow is a superposition of cyclic flows.

Now consider an arbitrary map colored in four colors, I, II, III, and IV, such that IV does not appear on the boundary. Then by the same procedure given in the above example, it is seen that there is a conservative flow in the edges such that the current strength in each edge is 1, 2, or 3 amperes.

Conversely, suppose that there is a conservative flow through the edges of a map such that the current strength in each edge is 1, 2, or 3 amperes. Now it is an elementary theorem of network theory that any conservative flow can be achieved by a superposition of cyclic currents flowing clockwise around the edges of the regions. Then the boundary regions have cyclic values ± 1, ± 2, ± 3 because the cyclic current must be equal to the boundary edge current. Next, observe that neighboring regions must have cyclic currents differing by an integer. Thus by moving from region to region it is seen that all cyclic currents are integer-valued.

Next, reduce the cyclic currents mod 4. Then we obtain a coloration of the map in colors I, II, III, and IV. Since the edge currents are not congruent to zero mod 4, it follows that adjacent regions have different color. Moreover, the boundary regions are not colored IV. This proves the following theorem:

A planar map can be colored in four colors if and only if there is a conservative flow through the edge network such that the current through an edge has strength 1, 2, *or* 3 *amperes.*

To extend these ideas to networks which are not planar, I made the following conjecture: consider a network such that every branch (edge) is on some circuit. Then there is a conservative flow such that the current through a branch has strength 1, 2, 3, or 4 amperes. Similar ideas were independently developed by W. T. Tutte.

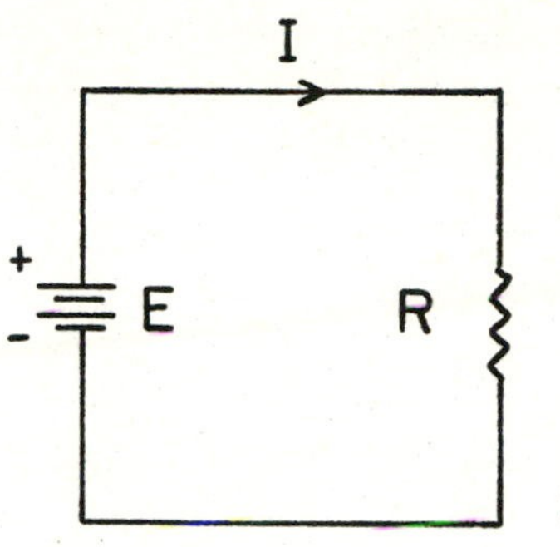

FIG. 2. A simple circuit.

2. SERIES-PARALLEL NETWORKS

Shown in Figure 2 is a simple circuit containing a battery of voltage E and a resistor of resistance R ohms. Then the current I flowing in the circuit is determined by the relation

$$E/I = R > 0. \tag{1}$$

This is Ohm's law.

Shown in Figure 3 are two resistors connected in series. One resistor has resistance A ohms and the other has resistance B ohms. Then the joint resistance R between terminals 1 and 2 is given by the formula

$$R = A + B. \tag{2}$$

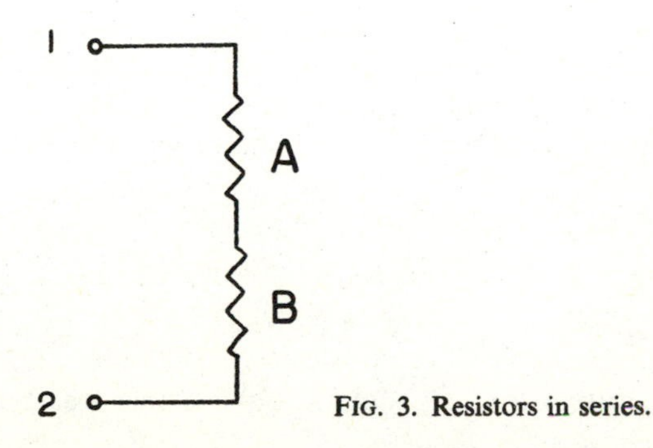

FIG. 3. Resistors in series.

On the other hand, the two resistors could be connected in parallel as shown in Figure 4. Conductance is the reciprocal of resistance and conductances add in the parallel connection so

$$R^{-1} = A^{-1} + B^{-1}.$$

Solving for R gives

$$R = AB/(A + B) \tag{3}$$

and this is the formula for the joint resistance R of two resistors in parallel.

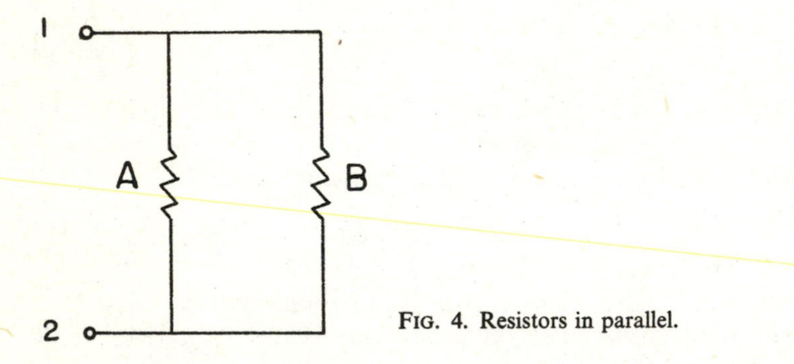

FIG. 4. Resistors in parallel.

To have a convenient short notation for the joint resistance of resistors connected in parallel let

$$A : B = AB/(A + B). \tag{4}$$

Then $A : B$ may be regarded as a new operation termed *parallel addition* [17]. Parallel addition is defined for any nonnegative numbers. The network model shows that parallel addition is commutative and associative. Moreover, multiplication is distributive over this operation.

Consider now an algebraic expression in the operations (+) and (:) operating on positive numbers A, B, C, etc. An example is

$$R = A + B : (C + D : E). \tag{5}$$

To give a network interpretation of such a polynomial, read

$A + B$ as "A series B" and $A : B$ as "A parallel B"; then it is clear that the expression (5) is the joint resistance of the network shown in Figure 5. Networks obtained from such polynomials are termed *series-parallel connections*.

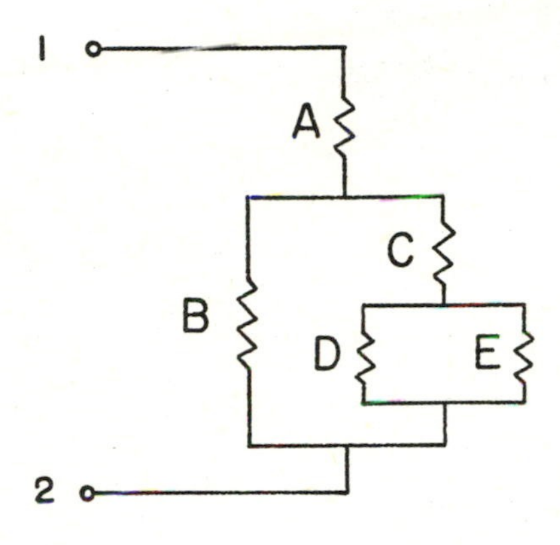

FIG. 5. A series-parallel connection.

Not every network is a series-parallel connection. In particular, it can be checked that the Wheatstone bridge connection of Figure 6 is not a series-parallel connection. In fact, it follows from an analysis given in [12] that a network is a series-parallel connection if and only if there is no embedded network having the Wheatstone bridge configuration. Another simple characterization of series-parallel connection has been given by Riordan and Shannon [23].

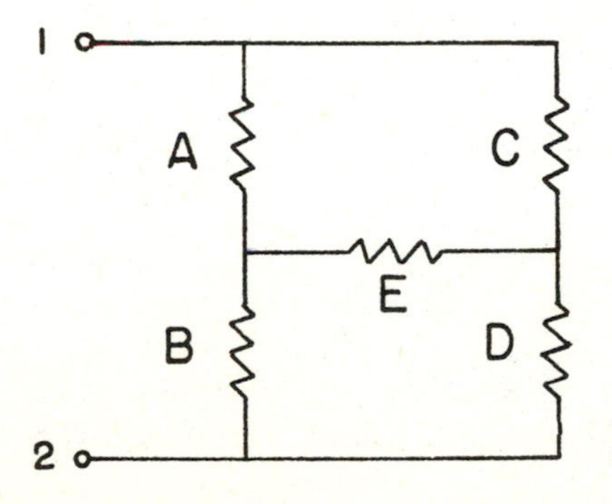

FIG. 6. The Wheatstone bridge connection.

According to a principle stated by Rayleigh [22] the current flow through a network may be described as taking the paths of least resistance. Alfred Lehman [18] used Rayleigh's principle to derive an interesting inequality termed *the series-parallel inequality*. He considered a network such as that shown in Figure 7. Then the joint resistance when the switch S is open is

$$R_\infty = (A + B) : (C + D).$$

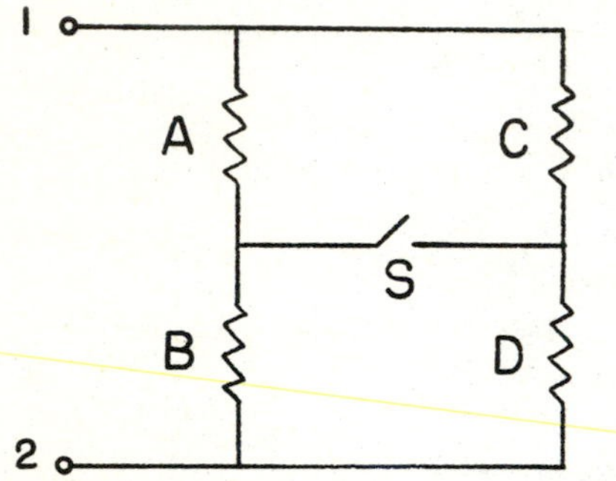

FIG. 7. Lehman's connection.

On the other hand, when the switch S is closed the joint resistance is

$$R_0 = A : C + B : D.$$

But the current takes the paths of least resistance and there is less constraint with the switch closed so

$$A : C + B : D \leqslant (A + B) : (C + D). \tag{6}$$

This is the series-parallel inequality and in ordinary algebra it is expressed as

$$\frac{AC}{A + C} + \frac{BD}{B + D} \leqslant \frac{(A + B)(C + D)}{A + B + C + D}.$$

It is worth noting that Lehman's connection corresponds to replacing the resistor E in the Wheatstone bridge connection with a switch. Let R_E denote the joint resistance of Wheatstone's bridge. Then the following is a generalization of Lehman's inequality:

$$R_0 \leqslant R_E \leqslant R_\infty. \tag{7}$$

This also is a consequence of Rayleigh's principle. The inequality on the right side of (7) is obtained by putting the resistor E in series with the switch. The inequality on the left is obtained by putting the resistor E in parallel with the switch.

3. THE PARALLEL ADDITION OF MATRICES

The various relationships just described become more interesting and suggestive when the scalar formulation of Ohm's law is replaced by a vector formulation. For example, Figure 8 depicts a resistor box with two pairs of terminals. The first pair is in circuit 1, denoted by a solid line, and the second pair is in circuit 2, denoted by a dashed line. Then the currents and voltages in these circuits satisfy equations of the form

$$E_1 = R_{11}I_1 + R_{12}I_2, \qquad E_2 = R_{21}I_1 + R_{22}I_2. \tag{8a}$$

In vector form, these equations can be written as

$$\mathbf{E} = R\mathbf{I}. \tag{8b}$$

If the box just contains interconnected resistors, then it results that R is a symmetric matrix. Moreover, by the conservation of energy it follows that R is positive semidefinite. Therefore, in what follows, an arbitrary symmetric positive semidefinite matrix R shall be termed a *resistance matrix*. For an appropriate generalization where R is not symmetric, see [13].

Resistance boxes may be added in series as is shown on the left side of Figure 9. If A and B are the resistance matrices of the two boxes, then the joint resistance matrix R is given by

$$R = A + B. \tag{9}$$

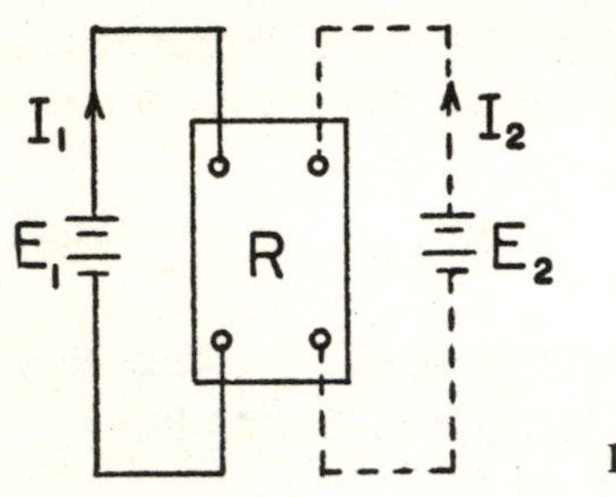

FIG. 8. A resistor box.

In other words, series connection of resistance boxes corresponds to addition of their matrices. (This assumes that the current I_1 in the first circuit of box A is the same as the current in the first circuit of box B. This can be achieved by use of isolation transformers.) The right side of Figure 9 gives an abbreviated symbolism for the series addition of resistor boxes.

Note that any current vector $\boldsymbol{I}$ can be the input to a resistor box.

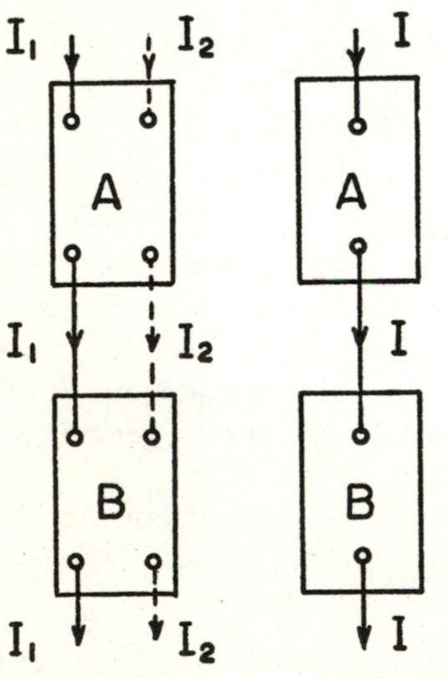

FIG. 9. Series addition of resistor boxes.

However, not every voltage vector $\boldsymbol{E}$ can be an output if R is a singular matrix. In any case, the following well-known theorem relates the range spaces of semidefinite matrices:

$$\text{Range}(A + B) = \text{Range}\, A + \text{Range}\, B. \tag{10}$$

It is equally possible to connect the resistor boxes in parallel as shown on the left side of Figure 10. The right side gives the symbolic diagram. First, suppose A and B are nonsingular, then

$$R^{-1} = A^{-1} + B^{-1}$$

where R is the joint resistance matrix of the parallel connection. Solving for R gives

$$R = A(A + B)^{-1}B. \tag{11}$$

Again it is convenient to have a short notation for the operation on the right side of (11), so let

$$A : B = A(A + B)^{-1}B \tag{12}$$

define *parallel addition* of matrices A and B. Various properties of

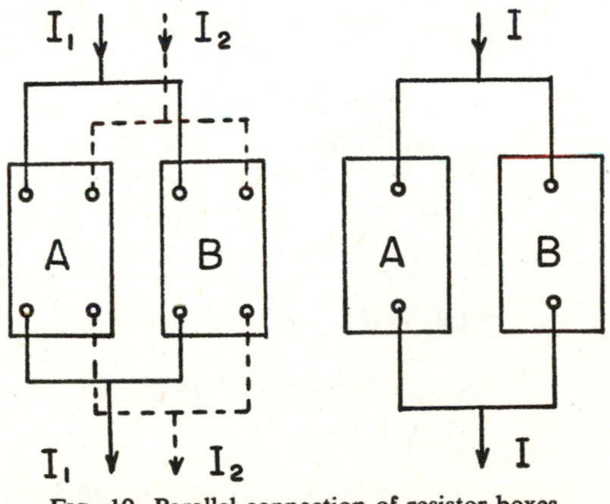

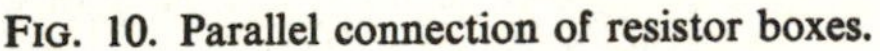
FIG. 10. Parallel connection of resistor boxes.

this new operation were developed in a joint paper with William N. Anderson [2]. Some of these properties will now be described. Other developments of parallel addition are given in [26], [37] and [39].

First, note that relation (10) shows that Range $(A + B) \supset$ Range B for semidefinite matrices. Hence $(A + B)^{-1}B$ is then well defined. This shows that the operation $A : B$ is defined for any pair of positive semidefinite matrices.

By virtue of the network model, we expect that the parallel sum is both commutative and associative. A direct proof of commutativity follows from the following manipulations:

$$A : B = (A + B - B)(A + B)^{-1}B = B - B(A + B)^{-1}B,$$

$$B : A = B(A + B)^{-1}(A + B - B) = B - B(A + B)^{-1}B.$$

This proves that

$$A : B = B(A + B)^{-1}A. \tag{13}$$

It is obvious from relation (12) that Range $(A:B) \subset$ Range A. Likewise, relation (13) shows that Range $(A:B) \subset$ Range B. Further analysis gives that actually

$$\text{Range}\,(A : B) = \text{Range}\,A \cap \text{Range}\,B. \tag{14}$$

Relations (10) and (14) reveal a remarkable duality between series addition and parallel addition of matrices.

To give an application of the above duality principle the networks shown in Figure 11a are now analyzed. Clearly, the joint resistance matrix of the first network is

$$R_1 = (A + B) : (B + C) : (C + A).$$

If α, β, and γ are the range spaces of A, B, and C respectively, it follows from (10) and (14) that

$$\text{Range}\,R_1 = (\alpha + \beta) \cap (\beta + \gamma) \cap (\gamma + \alpha).$$

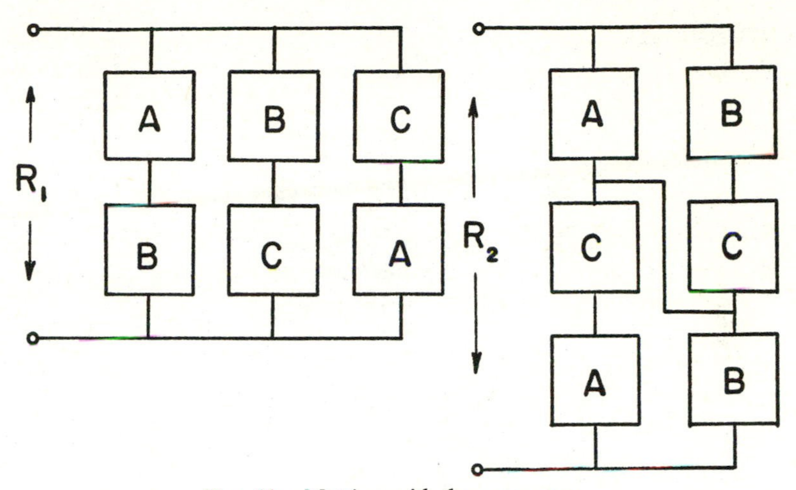

FIG. 11a. Matrices with the same range.

On the other hand, the joint resistance matrix of the second network is

$$R_2 = A : (B + C) + B : (A + C).$$

Therefore, the range of this matrix is Range $R_2 = \alpha \cap (\beta + \gamma) + \beta \cap (\gamma + \alpha)$. The subspaces of a vector space form a modular lattice. There are various identities which hold on a modular lattice. In particular,

$$(\alpha + \beta) \cap (\beta + \gamma) \cap (\gamma + \alpha) \equiv \alpha \cap (\beta + \gamma) + \beta \cap (\alpha + \gamma), \tag{15}$$

as is shown by Birkhoff [3]. This proves

$$\text{Range } R_1 = \text{Range } R_2. \tag{16}$$

The reader will see that there are various analogous procedures for constructing networks with the same range.

The parallel sum operation is found to satisfy various inequalities. Thus the norm, trace, and determinant satisfy the following inequalities:

$$\|A : B\| \leqslant \|A\| : \|B\|, \tag{17}$$

$$\mathrm{tr}(A : B) \leqslant (\mathrm{tr}\, A) : (\mathrm{tr}\, B), \tag{18}$$

$$\det(A : B) \leqslant (\det A) : (\det B). \tag{19}$$

Here the notation (:) on the right side of these inequalities denotes the scalar parallel operation. These inequalities give further manifestations of the duality between series and parallel addition.

The network connections used by Lehman to obtain the series-parallel inequality can be extended to resistor boxes. It is then found that

$$A : C + B : D \leqslant (A + B) : (C + D) \tag{20}$$

for positive semidefinite matrices A, B, C, and D. Here $A \leqslant B$ means that $B - A$ is positive semidefinite.

The scalar inequality (7) refers to the Wheatstone bridge and is a generalization of the series-parallel inequality. Presumably (7) can be extended to matrices; however, this is an open question.

Another type of connection of resistor boxes is termed the hybrid connection. In the hybrid connection some circuits are put in series and some are put in parallel. Such a connection is shown in Figure 11b. By use of the hybrid connection an elegant solution of the network synthesis problem was found [25].

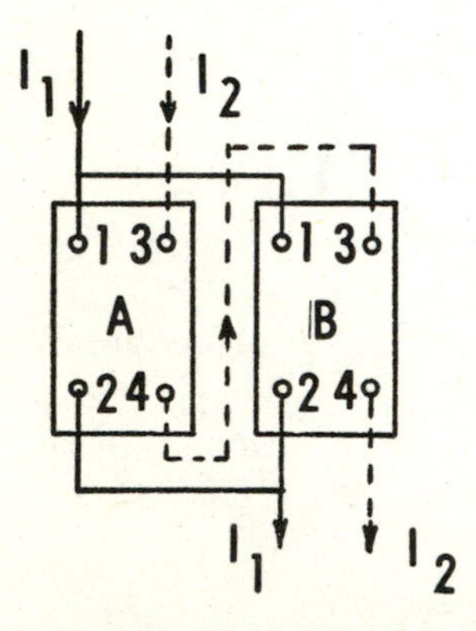

FIG. 11b. Hybrid connection of resistor boxes.

The joint resistance matrix R of the hybrid connection may be termed *hybrid addition* of matrices A and B. Some recent studies by George Trapp and the writer have revealed several properties of hybrid addition. In particular, the series-parallel inequality (20) is valid if $A : B$ now denotes hybrid addition of A and B [34].

4. THE BOTT-DUFFIN DUALITY ANALYSIS

It has long been known that many theorems about electrical networks have companion theorems obtained by interchanging current and voltage variables and replacing resistance by conductance. A theory of this electrical duality was developed in collaboration with Raoul Bott [4]. To explain this approach it suffices to consider a simple directed network such as shown in Figure 12. This network has six branches, so let $I_1, I_2, \ldots, I_6$ be the currents flowing in the directed branches. Then Kirchhoff's current law states that the total current entering a node vanishes. For example, at the node where branches (1, 2, 3) meet

$$I_1 - I_2 - I_3 = 0, \text{ etc.} \tag{21}$$

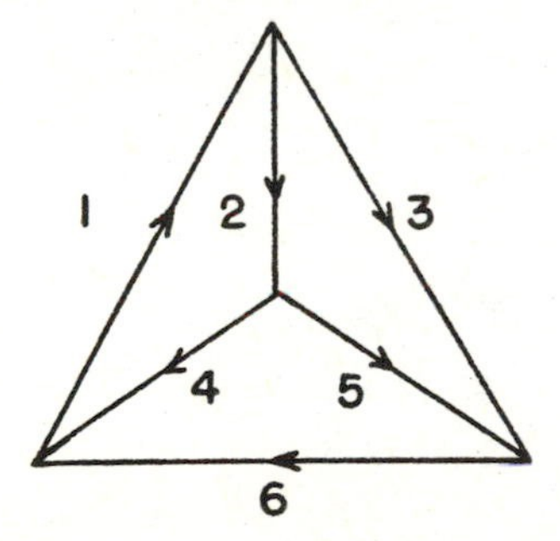

FIG. 12. A directed network.

Let $V_1, V_2, \ldots, V_6$ be voltage drops across the branches. Kirchhoff's voltage law states that the voltage sum around any circuit vanishes. For example, branches (1, 2, 4) form a circuit so

$$V_1 + V_2 + V_4 = 0, \text{ etc.} \tag{22}$$

It was observed by Herman Weyl [24] that the two laws of Kirchhoff cause current and voltage to be orthogonal, i.e.

$$\sum_{j=1}^{6} I_j V_j = 0. \tag{23}$$

This holds for any current flow I satisfying Kirchhoff's first law and any voltage V satisfying Kirchhoff's second law. For an application of (23) see [8].

Weyl's theorem (23) was the inspiration for our general duality analysis, but we changed (23) from a theorem to a postulate. Thus consider a six-dimensional Euclidean space E. Let K be the subspace of E corresponding to vectors $V = (V_1, V_2, \ldots, V_6)$ which satisfy Kirchhoff's voltage law. Then let P be the perpendicular projection operator from E into K. Thus P is a symmetric matrix such that $P^2 = P$. Let $P' = I - P$, then P' is also a perpendicular projection matrix and $PP' = 0$. Moreover, P' projects E into K', the orthogonal complementary subspace of K. By Weyl's theorem, it follows that K' is the space corresponding to vectors satisfying Kirchhoff's current law.

To relate the perpendicular projection matrix P to electrical properties, let G be a diagonal matrix whose diagonal elements $g_1, g_2, \ldots, g_6$ are positive numbers giving the conductances of the six branches. Then the *discriminant* D is defined as

$$D = \det(GP + P'). \tag{24a}$$

Thus D is a multilinear form in $g_1, \ldots, g_6$:

$$D = g_1 g_2 g_3 + \cdots + g_2 g_4 g_5. \tag{24b}$$

Each term of D is a product of branches which form a tree of the network.

The *transfer matrix* T is defined as

$$T = P(GP + P')^{-1}. \tag{25}$$

This matrix has the following physical significance: suppose that a current source of strength I is inserted in branch 6. Then the voltage across branch 2 is $V_2 = T_{26}I$. In fact *all electrical properties of the network are given by the transfer matrix T.*

Electrical duality comes about by defining G' to be a diagonal matrix whose diagonal elements $g_1', g_2', \ldots, g_6'$ are the resistances of the branches. Then the *dual discriminant* is defined as

$$D' = \det(G'P' + P). \tag{26a}$$

Thus D' has the form

$$D' = g_4' g_5' g_6' + \cdots + g_1' g_3' g_6'. \tag{26b}$$

Each term of D' is a product of branches whose complement is a tree. The *dual transfer matrix* is

$$T' = P'(G'P' + P)^{-1}. \tag{27}$$

This matrix has the following physical significance: suppose that a battery of voltage E is inserted in branch 6, then the current in branch 2 is $I_2 = T_{26}'E$.

We term the correspondence between primed and unprimed symbols *electrical duality*. Moreover, there are various relations between the primal and dual symbols. For example,

$$G'T' + TG = 1. \tag{28}$$

A beautiful algebraic structure develops from the above concepts. Moreover, two far-reaching generalizations are feasible: (i) P can be chosen to be an arbitrary perpendicular projection matrix. (ii) G can be taken to be an arbitrary matrix. In this generalization, T is termed the *constrained inverse* of G. The constrained inverse exists if and only if the discriminant $D \neq 0$. These generalizations also have electrical and mechanical interpretations. See also [36] and [39].

5. HOW TO USE THE WANG ALGEBRA

K. T. Wang managed an electrical power plant and in his spare time he sought simple rules for solving the network equations.

These rules appear to have a connection with the Bott-Duffin analysis. To make this connection, Bott and I restated his rules as three postulates for an algebra:

$$xy = yx, \tag{i}$$

$$x + x = 0, \tag{ii}$$

$$xx = 0. \tag{iii}$$

Here x and y are arbitrary elements of the algebra [5], [10].

To see how to apply the Wang algebra, consider the network shown in Figure 13. Let the branch conductances a, b, c, d, and e be regarded as independent generators of the Wang algebra. A star element of the algebra is defined to be the sum of branches meeting at a node. Thus the star element at node 3 is $(a + b + c)$. To find the discriminant, the rule is to carry out the Wang product of all stars except one. Then omitting the star at node 2 gives

$$D = a(a + b + c)(d + e + c) = (ab + ac)(d + e + c).$$

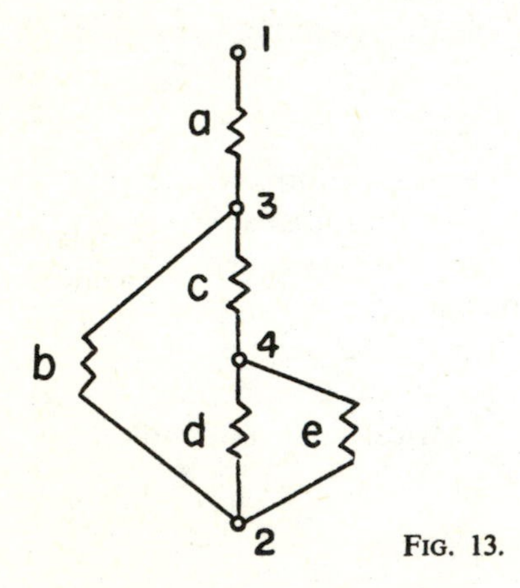

FIG. 13.

Thus the discriminant of the network is

$$D = abd + abe + acd + ace + abc. \tag{29}$$

It can be readily checked that the terms of (29) give all the trees of the network of Figure 13.

To find the joint resistance R between nodes 1 and 2, the rule is to write R as a fraction. The denominator of the fraction is the discriminant D and the numcrator N is the product of all stars except those at 1 and 2. Thus

$$R = \frac{N}{D} = \frac{ad + ae + bd + be + cd + ce + ac + bc}{abd + abe + acd + ace + abc} . \quad (30)$$

There are also simple rules for calculating the transfer matrices T and T'.

The network shown in Figure 13 is a series-parallel connection; in fact, it is the same connection as is shown in Figure 5. Thus the joint resistance R could also be calculated by the series-parallel formula (5). However, the Wang rules apply even if the network is not of series-parallel type.

A proof of the Wang rules was made by first observing that the Wang algebra is the Grassmann algebra when the coefficient field is the integers mod 2. However, the Grassmann algebra gave a more general system of calculation and one which could be directly related to the Bott-Duffin analysis.

For example, let the symbols a, b, c, and d now be regarded as independent vectors of a real vector space E_4. Then the vectors $(a + d)$, $(a + b + c)$, $(d + c - a)$ form a basis of a subspace S but they are not stars (or circuits) of any network. The Grassmann algebra consists of the vectors of E_4 together with outer products formed by the law $xy = -yx$ for any two vectors x and y. Then the outer product π associated with S is

$$\begin{aligned}\pi &= (a + d)(a + b + c)(d + c - a) \\ &= abd + acd + abc + dac + dbc - dba - dca \\ &= 2abd + 3acd + abc + dbc.\end{aligned}$$

Then according to the Bott-Duffin rule the coefficients of the discriminant are the square of the coefficients of the outer product; thus

$$D = 4abd + 9acd + abc + dbc.$$

Now a, b, c, and d denote real numbers.

6. WHAT IS A REGULAR MATROID?

There are ideal electrical networks which do not obey Kirchhoff's law. For example, Figure 14 shows a double triangle network linked with a magnetic ring of zero magnetic resistance. Then by Ampere's law, no electric current may link the ring. This imposes a constraint in addition to Kirchhoff's current law. Nevertheless, the Bott-Duffin analysis applies without change. More remarkable is the fact that the Wang algebra also works. For example, the dual discriminant is given by the Wang product $D' = \alpha\beta\gamma$ where α, β, and γ are the three square circuits indicated in Figure 14.

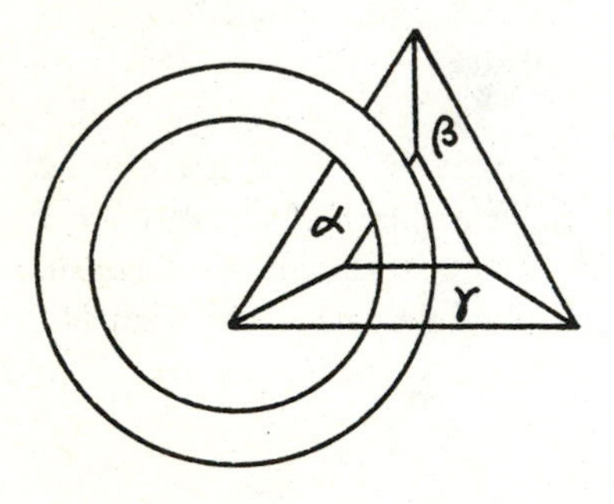

FIG. 14. A linked network.

This raises the following question: what characterizes subspaces such that the discriminant can be calculated by Wang's short cut method? Let us term such subspaces *quasi-Kirchhoffian*. Then an answer to the question is given by the following statement which Bott and I proved in 1951 [5], [10]:

QK THEOREM. *Let S be an m-dimensional subspace of n-dimensional real vector space E_n. Let C be the set of those vectors of S whose components are $+1$, -1, or 0. Then S is quasi-Kirchhoffian if and only if C is an m-dimensional vector space under addition* mod 2.

To understand the application of this theorem, first suppose that S is the subspace defined by Kirchhoff's current law. Then C is taken to be the set of current flows of unit strength. Thus a vector C_1 of C is composed of one or more non-overlapping circuits. Moreover, it is an easy exercise in graph theory to show that then S satisfies the QK theorem.

The proof of the QK theorem in the general case proceeds by treating C as a matrix of column vectors and showing that it is *totally unimodular*. A matrix is totally unimodular if each minor determinant has the value 1, -1, or 0. The concept of total unimodularity plays an important role in integer linear programming [40]. Thus Wang algebra and integer linear programming are seen to be related.

Wang algebra leads directly to a generalization of graph theory. To see this, term the vectors of C *quasi-circuits*. Then the QK theorem has the following easy corollary:

QK COROLLARY: *Let C_1 be a quasi-circuit whose first and second components do not vanish. Let C_2 be a quasi-circuit whose first component does not vanish but whose second component does vanish. Then there is a quasi-circuit C_3 whose first component vanishes but whose second component does not vanish. Moreover, if the i-th components of C_1 and C_2 vanish, then the i-th component of C_3 must vanish.*

The QK corollary is obviously true if C_1 and C_2 are circuits of a graph. In fact, it is this property which Whitney used as a postulate to characterize a matroid [41].

Further analysis reveals that the QK theorem is a necessary and sufficient characterization of the so-called *regular matroids* studied by Tutte [42]. An elegant analysis of the relation of regular matroids to network theory is given by Minty [20].

7. SQUARING THE SQUARE

Is it possible to divide a square into squares no two of which are equal? This puzzle resisted attack for years but fell before a massive assault by Brooks, Smith, Stone, and Tutte. Making use of an electric network model, these authors developed an example in which a square was divided into 26 smaller squares [6]. The minimum number is not known.

The network they employed may be described as a lumped network equivalent to a distributed network. To understand this correspondence, consider a situation when a rectangle is divided into smaller rectangles such as shown in Figure 15. Suppose that the rectangle is constructed out of a thin conducting plate such that a unit square has a resistance of one ohm between opposite sides. Since resistance is proportional to length and inversely proportional to width, it follows that a square of any size has resistance of one ohm. Thus if E is the height of the rectangle and I its width, then the resistance from top to bottom is $R = E/I$.

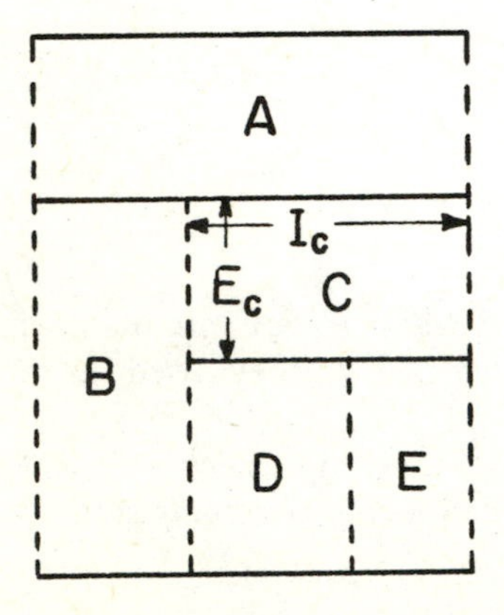

FIG. 15. A rectangular network.

The streamlines of current flow in the rectangle are vertical lines. Thus, the flow will not be changed if cuts are made along the vertical dashed lines separating the rectangles marked B, C, D, E. The equipotentials are horizontal lines. Thus the flow would not be changed if perfectly conducting bus bars are placed along the solid lines forming upper and lower sides of the rectangles. By virtue of these observations the plate may be regarded as a lumped network having lumped resistors of value A, B, C, D, and E ohms. These resistances are determined by the dimensions. Thus if I_C is the width and E_C is the height of the rectangle marked C, then $C = E_C/I_C$ etc.

Clearly the lumped network is the same as the series parallel connection shown in Figure 5. Thus the resistance of the plate is

$$R = A + B : (C + D : E) \tag{31}$$

because the cuts and bus bars do not change resistance. This example of a rectangular network suggests the following conjecture:

Every series-parallel connection has an equivalent division of a rectangle into rectangles.

8. RAYLEIGH'S RECIPROCAL RELATION

Again consider a conducting plate having resistance of 1 ohm between opposite sides of a unit square. Figure 16 shows a curvilinear quadrilateral plate with sides 1, 2, 3, and 4. The sides 3 and 4 are insulated but sides 1 and 2 are connected to perfectly conducting bus bars (denoted by heavy lines). Let R_{12} be the joint resistance between sides 1 and 2. Next, let sides 1 and 2 be insulated and let bus bars be connected to sides 3 and 4. If R_{34} is the joint resistance in this dual situation, then

$$R_{12}R_{34} = 1. \tag{32}$$

This is Rayleigh's reciprocal relation [21].

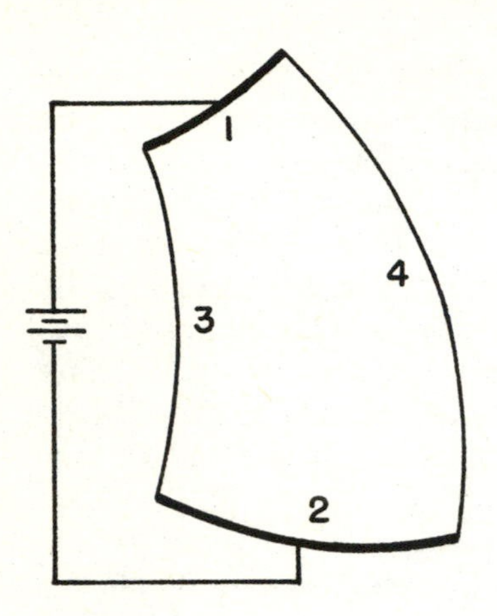

FIG. 16. A curvilinear quadrilateral.

To prove (32) draw the equipotential lines and the streamlines. Since the potential $u(x, y)$ is harmonic, the streamlines are the equipotential lines of the conjugate harmonic function $v(x, y)$. It follows from the Cauchy-Riemann equations that the two sets of equipotentials are orthogonal and divide the region into curvilinear squares such as is shown in Figure 17. This breaks the flow up into channels. One of these channels is denoted by crosshatching in Figure 17. This channel is a series of 7 squares so the total resistance is 7 ohms. It is seen that there are 4 channels in parallel so this gives $R_{12} = 7/4$ ohms. Now consider the conjugate problem; then Figure 17 again applies, but the equipotentials and streamlines interchange roles. Thus a channel from side 3 to side 4 is a series of 4 squares and so the channel resistance is 4 ohms. There are 7 channels so the total resistance is $R_{34} = 4/7$ ohms. This is the proof of (32) given by Rayleigh.

A surprising consequence of the Rayleigh relation arises when the quadrilateral region has bilateral symmetry as in Figure 18. By symmetry $R_{12} = R_{34}$. Hence by Rayleigh's reciprocal relation $R_{12} = 1$ ohm.

These considerations raise the question of a lumped network analog of Rayleigh's reciprocal relation [9], [11]. To answer this question consider Figure 19. A planar network is shown in solid lines with two distinguished nodes, 1 and 2. Another planar network is shown in dashed lines, and it has two distinguished

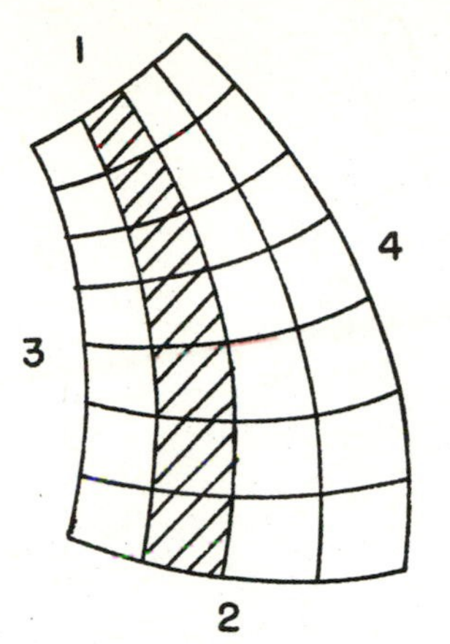

FIG. 17. Conjugate functions.

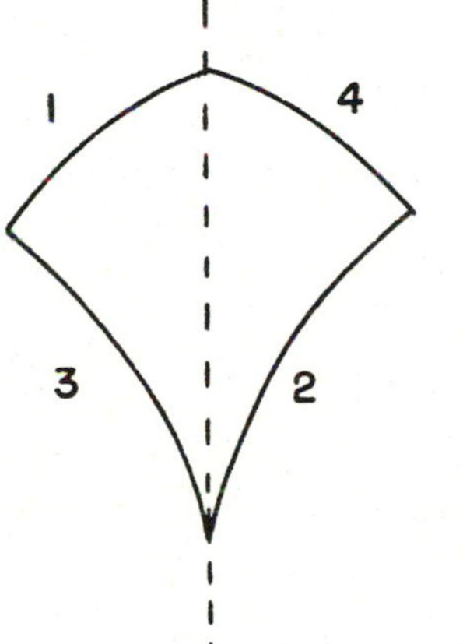

FIG. 18. Self-dual conductor.

nodes, 3 and 4. These networks are termed *dual* because of the following properties: (i) Crossing branches give a one-to-one correspondence between the networks. (ii) A region of one of the networks contains one and only one of the nodes of the other network. (iii) The distinguished nodes are on the boundary and are not in a region. (iv) If branches cross, the resistances r and r^* are required to satisfy

$$rr^* = 1. \tag{33}$$

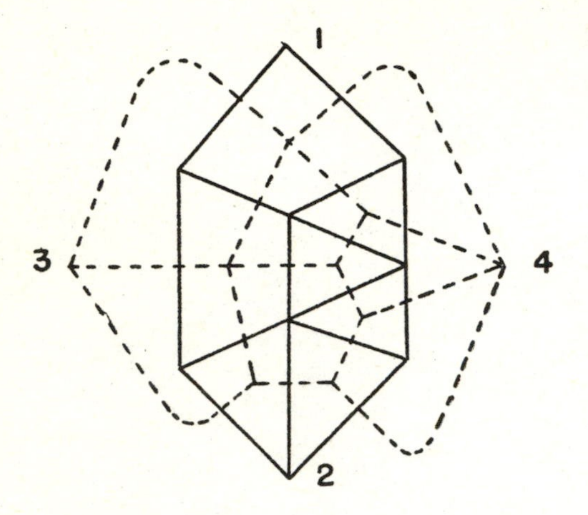

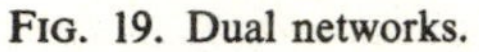
FIG. 19. Dual networks.

Under these hypotheses it follows that the reciprocal relation (32) holds for the joint resistance of the two networks.

A proof of (32) can be given by noting that Kirchhoff's current law for the primal network defines the same constraint as Kirchhoff's voltage law for the dual network. It follows immediately from the Bott-Duffin analysis that the resistance of the primal network is equal to the conductance of the dual network.

9. UPPER BOUND AND LOWER BOUND NETWORKS

The potential $u(x, y)$ of a conducting plate satisfies Laplace's equation

$$\partial^2 u/\partial x^2 + \partial^2 u/\partial y^2 = 0.$$

To solve potential problems on a computer, Laplace's equation is approximated by the difference equation

$$u(x, y) = [u(x+h, y) + u(x-h, y) + u(x, y+h) + u(x, y-h)]/4. \tag{34}$$

Physically, this corresponds to replacing the conducting plate by a wire screen having square meshes of side h. Then Equation (34) states that the potential at a node is the mean of the potentials at the four neighboring nodes. This is clearly a consequence of the laws of Ohm and Kirchhoff.

It is important to know the nature of the error in replacing the plate by the wire screen. As an approach to this problem, consider the rectangular plate shown in Figure 20. The plate is 4 cms. high and 3 cms. wide so the resistance between the top edge and bottom edge is 4/3 ohms. This suggests that the screen wire should have a resistance of one ohm per centimeter. First suppose that the screen is placed as shown by the dashed lines. Then the horizontal wires carry no current but the current flows through three vertical wires, each having a resistance of 4. The total resistance is 4/3 ohms so there is no error in using the dashed network. Next consider the network indicated by solid lines. Now there are 4 vertical wires and this poses a problem. However, if it is supposed that the wires on the boundary have resistance of 2 ohms per cm., then the correct joint resistance of 4/3 ohms is again obtained.

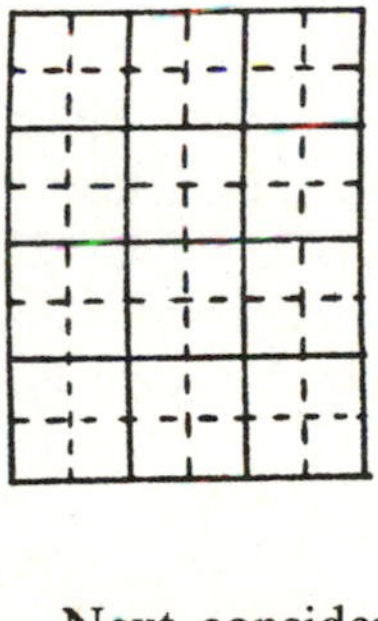

FIG. 20. Plate and screen.

Next consider an arbitrary region made up of squares. Such a plate is shown in Figure 21. Let R be the joint resistance from edge 1 to edge 2 when the other edges are insulated. Two screen networks are shown in Figure 21. The resistance of the square sides are to be 1 ohm inside and 2 ohms on the boundary. We term the solid lines the *upper* network. Let R^u be the joint resistance of the upper network between edges 1 and 2. We term the dashed lines the *lower* network. Let R^L be the joint resistance of the lower network. Then the following inequality maintains:

$$R^u \leqslant R \leqslant R^L. \tag{35}$$

Thus the upper network furnishes an upper bound to the conductance and the lower network furnishes a lower bound to conductance [9].

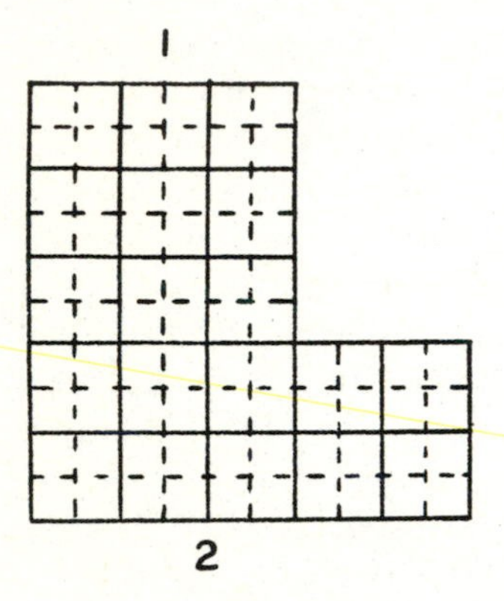

FIG. 21. An *L*-shaped plate.

The proof of (35) is obtained as a special case of a more general theorem in which a polygonal region is triangulated in an arbitrary fashion. Thus consider the polygonal plate shown in Figure 22. The triangulation of this polygon is denoted by solid lines. The solid lines are termed the upper network. An interior line p of this network is given resistance

$$r = 2/(\cot \alpha + \cot \beta) \tag{36}$$

where α and β are the angles opposite the line p as shown in Figure 22. If p is a boundary line, then the term $\cot \alpha$ would be omitted in this formula.

Let $u(x, y)$ be the potential function for the plate problem. Let $w(x, y)$ be a function which is linear in each triangle but which is

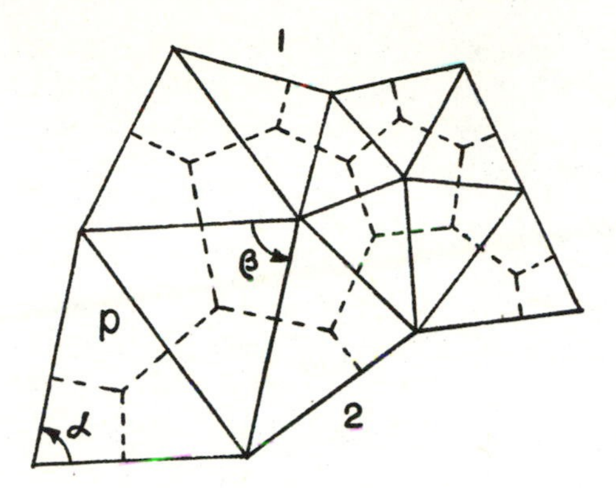

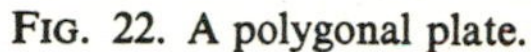
FIG. 22. A polygonal plate.

continuous over the plate region. Then by Dirichlet's principle

$$\iint\left[\left(\frac{\partial u}{\partial x}\right)^2+\left(\frac{\partial u}{\partial y}\right)^2\right]dx\,dy \leqslant \iint\left[\left(\frac{\partial w}{\partial x}\right)^2+\left(\frac{\partial w}{\partial y}\right)^2\right]dx\,dy$$

provided $u = w$ on both boundary segment 1 and boundary segment 2. Then by making certain transformations, this inequality is found to be equivalent to

$$1/R_{12} \leqslant 1/R_{12}^{u} \tag{37}$$

where R_{12} is the joint resistance of the plate and where R_{12}^{u} is the joint resistance of the upper network. Thus the upper network furnishes an upper bound to conductance.

To obtain a lower bound for conductance we may use Rayleigh's reciprocal relation $R_{12} = 1/R_{34}$. Here 3 and 4 denote boundary segments complementary to segments 1 and 2. But by formula (37) it follows that $1/R_{34} \leqslant 1/R_{34}^{u}$. Thus

$$R_{12}^{u} \leqslant R_{12} \leqslant 1/R_{34}^{u} \tag{38}$$

gives upper and lower bounds to R_{12}.

The *lower* network is defined to be the dual of the *upper* network. The lower network is shown by dashed lines in Figure 22.

The resistance of a dashed branch is the reciprocal of the resistance of a branch it crosses. It follows that the joint resistance R_{12}^L of the lower network satisfies the reciprocal relation $R_{12}^L = 1/R_{34}^u$. Substituting this in (38) gives

$$R_{12}^u \leqslant R_{12} \leqslant R_{12}^L. \tag{39}$$

This is the desired bounding relation. For further generalizations see [14] and [16].

If a region is covered by a square lattice, then it can be triangulated by inserting one diagonal in each square. It then results that (39) gives the bounding relation (35) stated for square lattices.

10. THE EXTREMAL LENGTH OF A NETWORK

Ahlfors and Beurling have introduced the concept of *extremal length* of a curvilinear quadrilateral [1]. The relation of this geometric concept to complex function theory has been developed at extreme length in the literature so we shall only give the definition. Consider the curvilinear quadrilateral shown in Figure 23. Then the extremal length is denoted by EL and is defined as

$$EL = \sup_{w} \inf_{P} \frac{\left(\int_P w\, ds\right)^2}{\int\int_G w^2\, dx\, dy}. \tag{40}$$

Here P is any path from side 1 to side 2 and $w(x, y)$ is any continuous nonnegative function defined over the region G.

To give an electrical interpretation of extremal length we imagine the quadrilateral G is a conducting plate of unit resistivity. Let R be the joint resistance between sides 1 and 2. Then $E = IR$ where E is the battery voltage and I is the current flow. The power input is $EI = E^2/R = I^2R$. On the other hand, if $w(x, y)$ is the

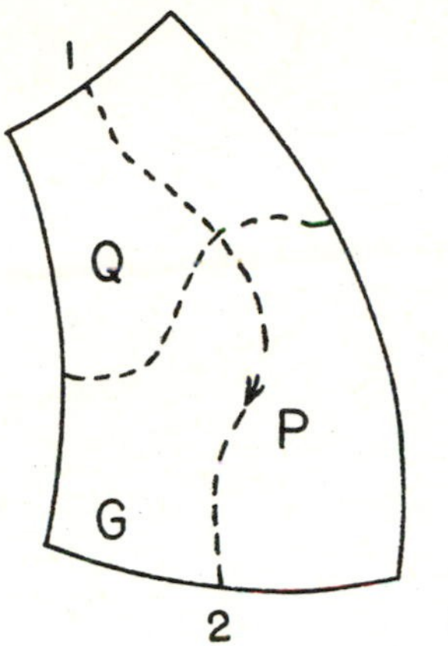

FIG. 23. A region G.

strength of the current density, the power dissipated in heat is w^2 per square centimeter. Thus by the conservation of energy

$$E^2/R = \int\int_G w^2 \, dx \, dy. \tag{41}$$

If P is a streamline from 1 to 2,

$$E = \int_P w \, ds. \tag{42}$$

Combining (41) and (42) gives

$$R = \frac{\left(\int_P w \, ds\right)^2}{\int\int_G w^2 \, dx \, dy}. \tag{43}$$

It is then possible to show that this choice of w and P is optimal. Thus the extremal length is simply the joint resistance.

The concept of extremal length can be extended to networks in the following way [11]: consider a network G with two distinguished nodes 1 and 2. A *path* P connecting nodes 1 and 2 is designated in Figure 24 by arrows. Let r_j denote the resistance of

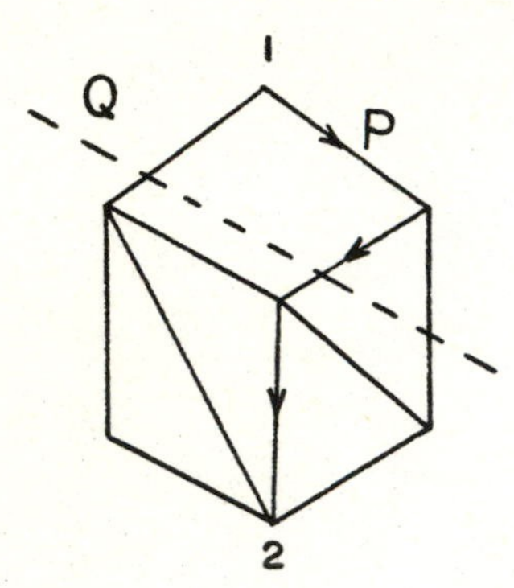

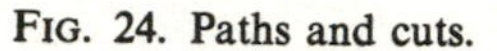
FIG. 24. Paths and cuts.

branch j. Then the extremal length is defined to be

$$EL = \max_{w} \min_{P} \frac{\left(\sum_{P} r_j w_j\right)^2}{\sum_{G} r_j w_j^2}. \tag{45}$$

Here w_j is an arbitrary nonnegative function defined on the branches. If w_j is actually the strength of the current in branch j, then $\Sigma_G r_j w_j^2$ is the power dissipated in heat. If the path P follows the direction of current flow from 1 to 2, then $\Sigma r_j w_j = E$, the battery voltage. Thus by the conservation of energy, the joint resistance satisfies

$$R = \frac{\left(\sum_{P} r_j w_j\right)^2}{\sum_{G} r_j w_j^2}. \tag{46}$$

Again it can be shown that this choice of P and w is optimal. Hence the extremal length of a network is equal to the joint resistance.

The network shown in Figure 24 is planar; however, the definition holds for general networks. Moreover, the formulation of the network concept of extremal length suggests a related concept termed *extremal width*. The extremal width is denoted by EW and is defined as

$$EW = \max_{w} \min_{Q} \frac{\left(\sum_{Q} w_j\right)^2}{\sum_{G} r_j w_j^2}. \tag{47}$$

Here Q denotes a cut and is defined as a set of branches which separate node 1 from node 2. A cut is indicated in Figure 24. If w_j is actually the current strength in branch j, then $\min \Sigma_Q w_j = I$ the total current. Also $\Sigma_G r_j w_j^2 = I^2 R$ hence

$$1/R = \frac{\left(\sum_{Q} w_j\right)^2}{\sum_{G} r_j w_j^2}. \tag{48}$$

This is actually the optimal solution. Thus the following identity holds:

$$(EL)(EW) = 1. \tag{49}$$

It is worth noting that the definition of joint resistance by means of extremal length makes no explicit appeal to either of Kirchhoff's laws.

If in relation (49) the maximization operations are omitted, then an inequality results. Writing v_j for $r_j w_j$ gives the network inequality

$$\sum_{G} v_j w_j \geqslant \left(\min_{P} \sum_{P} v_j\right)\left(\min_{Q} \sum_{Q} w_j\right). \tag{50}$$

This is termed the *width-length inequality* [11]. It holds for arbitrary $w_j \geqslant 0$ and $v_j \geqslant 0$. Here v_j is arbitrary because r_j is arbitrary. A special case of (50) had previously been found by Moore and Shannon [43]; a generalization is given by Lehman [19].

The fact that the width-length inequality holds for arbitrary networks suggests that an analogous relation holds for conducting bodies. Thus consider a topological image G of the cylinder. For example, Figure 23 can serve to illustrate this situation but G is now considered to be a solid body rather than a plate. Let the top surface of the "cylinder" be denoted by 1 and the bottom surface be denoted by 2. Thus P denotes a path in G from 1 to 2. Likewise Q represents a surface cutting the body in two parts such that 1 and 2 are not in the same part. Then the following width-length inequality was conjectured in [11]:

$$\int\int\int_G VW\,dx\,dy\,dz \geqslant \left(\inf_P \int_P V\,ds\right)\left(\inf_Q \int\int_Q W\,dA\right) \tag{51}$$

where $V \geqslant 0$ and $W \geqslant 0$ are arbitrary continuous functions. Recently this inequality has been proved by W. R. Derrick [7], [29].

11. DUAL PROGRAMS FOR OPTIMIZING HEAT TRANSFER

Again we are concerned with the joint conductance of a network, but now an economic constraint is introduced in addition to the physical constraints of Kirchhoff. The problem is expressed in the form of a program. It results that the solution is obtained by minimizing a norm over a vector space. This norm is of mixed type. That is, part of the norm comes from the economics and part comes from the physics.

Consider a network such as that shown in Figure 24, but suppose that the branches are divided into two sets, α and β. In set α the conductances $g_j \geqslant 0$ are fixed. In set β the conductances are allowed to vary but are subject to the constraint

$$\sum_\beta g_j \leqslant K, \qquad g_j \geqslant 0.$$

This constraint corresponds to limiting the total weight of the β branches. The problem is to distribute the material so as to maximize the joint conductance relative to terminals 1 and 2.

Let M denote the maximum joint conductance. Then M satisfies the following duality inequality [15], [28].

$$\underset{x \in S}{\|x\|_P^2} \geqslant M \geqslant \underset{y \in T}{\|y\|_D^{-2}}.$$

Here $\|x\|_P$ is a norm on the vector $(x_1, x_2, \ldots, x_n)$ defined as

$$\|x\|_P^2 = \sum_\alpha g_j x_j^2 + K \max_\beta x_j^2.$$

On the right, $\|y\|_D$ is a norm on the vector $(y_1, y_2, \ldots, y_n)$ defined as

$$\|y\|_D^2 = \sum_\alpha g_j^{-1} y_j^2 + K^{-1}\left(\sum_\beta |y_j|\right)^2.$$

The constraints on the vectors x and y are: S. $x_1, \ldots, x_n$ are a possible set of potential differences on the branches corresponding to unit potential difference across the input terminals. T. $y_1, \ldots, y_n$ are a possible set of branch currents corresponding to a unit flow of current at the input terminals. In other words, the x_j satisfy the Kirchhoff voltage law and the y_j satisfy the Kirchhoff current law.

It is seen from the above definition that the norm $\|x\|_P$ is a mixture of a Euclidean norm and a Tchebychef norm. Likewise the norm $\|y\|_D$ is a mixture of a Euclidean norm and an L_1 norm. These "mixed norms" would seem somewhat artificial if it were not for the fact that they express the essence of a natural problem of economics.

These inequalities for networks have analogies for continuous bodies. Thus Duffin and McLain [32] have thereby determined the most efficient design for a cooling fin. Bhargava and Duffin have given extensions with the L_p norm [28].

12. DUAL PROGRAMS FOR NONLINEAR NETWORKS

The networks and conducting bodies considered above are assumed to obey Ohm's law. In other words, there is a linear relation between current flow and potential difference. Now it is desired to treat nonlinear networks, and these will be discussed in the light of the Legendre transform.

Let $u(x)$ be a smooth convex function defined on an open region R of n-dimensional space E. Then each point x of R is mapped into a point y by the relation

$$y_j = \frac{\partial u}{\partial x_j} \qquad j = 1, 2, \ldots, n.$$

Let R^* be this map of R. Then the Legendre transform is defined as

$$v = u(x) - \sum_1^n x_j y_j.$$

It is easily shown that v is a single valued function of y in R^*.

The following duality inequality holds between a function u, its Legendre transform v and a constant M [31]:

$$\underset{x \in S}{u(x)} \geqslant M \geqslant \underset{y \in T}{v(y)}.$$

Here the constraints are:

$$\text{S.} \qquad x \in R \cap K,$$

$$\text{T.} \qquad y \in R^* \cap K'.$$

Here K is a subspace of E and K' is the orthogonal complementary subspace.

This simple duality inequality has many applications. In particular, it was used to treat certain geometric programs [31]. Another important application is to nonlinear networks [30]. Thus let a network have n branches. Then y_j represents current in branch j and x_j represents voltage drop across branch j. Then for each branch there is a functional relation of the form

$$y_j = g_j(x_j).$$

In a common situation, the branch conductances are such that $g_j(\ )$ is a continuous monotone increasing function. I have termed such networks "quasi-linear" while Minty has termed them "monotone" [35]. (In general $g_j(0) \neq 0$.)

Let the function $u(x)$ be defined as

$$u(x) = \sum_1^n G_j(x_j)$$

where

$$G_j = \int_0^x g_j(x)\, dx.$$

Let r_j be the inverse function to g_j, so $x_j = r_j(y_j)$. Then it is easy to check that

$$v(y) = \sum_1^n \left[G_j(r_j(y_j)) - y_j r_j(y_j) \right]$$

is the Legendre transform.

The subspace K is the Kirchhoff voltage space. The subspace K' is the Kirchhoff current space. The orthogonality of current and voltage is an important concept due to Weyl. Weyl's concept was further developed by Bott and Duffin [4, 5] and by Tellegen [38].

The primal program is the minimization of $u(x)$ under the constraint S. A minimum point is a desired equilibrium state of the system. The dual program is the maximization of $v(y)$ under the constraint T. If the primal program has a minimum value M, then the dual program has a maximum M. The value M has an important significance for linear networks but not for nonlinear networks.

13. TRIPARTITE GRAPHS WHICH JOIN NETWORKS

Section 3 concerned two resistor boxes A and B having n terminals. These resistor boxes were interconnected to form a new resistor box C also having n exterior terminals. The interconnections discussed were termed series, parallel and hybrid. Clearly, there are many other possible ways of interconnecting resistor boxes. For example, the cascade connection shown in Figure 25 is of considerable practical importance.

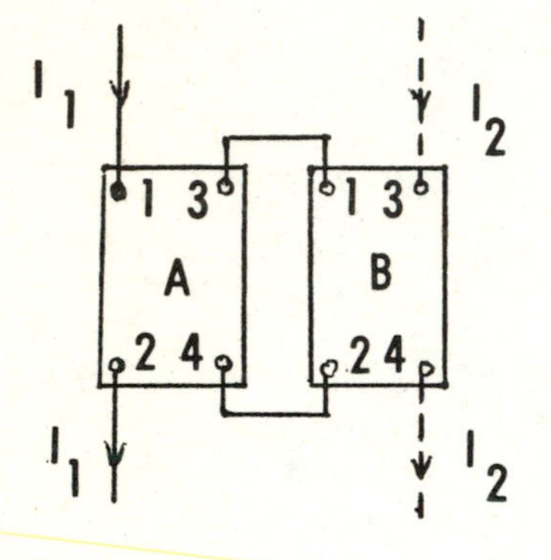

FIG. 25. Cascade connection of resistor boxes.

To give a general analysis of the interconnection of resistor boxes, Anderson, Trapp and I have introduced the concept of a "junctor". A junctor is a graph having $3n$ vertices divided into three equal sets. In other words, a junctor is a tripartite graph. The first set of vertices will be joined to resistor box A. The second set of vertices is joined to resistor box B. The third set will be the terminals for the new resistor box.

A desirable physical property is that the junctor itself should not restrict the possible current flows entering the boxes. Thus we forbid unconnected terminals or short circuited terminals. A precise set of requirements which insures this behavior follows.

A *junctor* is a graph whose set of vertices is divided into three equal classes, $A = \{a_i\}$, $B = \{b_i\}$, and $C = \{c_i\}$ such that

i. Each vertex is connected to some other vertex.

ii. No vertex is connected to another vertex of its own class.

iii. No vertex is connected to two distinct vertices of the same class.

A junctor is thus specified by the three incidence matrices K, L, and M, where

$$k_{ij} = \begin{cases} 1 \text{ if vertex } a_j \text{ is connected to vertex } c_i, \\ 0 \text{ otherwise,} \end{cases}$$

$$l_{ij} = \begin{cases} 1 \text{ if vertex } b_j \text{ is connected to vertex } c_i, \\ 0 \text{ otherwise,} \end{cases}$$

$$m_{ij} = \begin{cases} 1 \text{ if vertex } a_j \text{ is connected to vertex } b_i, \\ 0 \text{ otherwise.} \end{cases}$$

It follows from the definition that K, L, M each have the property that no row or column contains more than one 1. Alternatively, we may specify the junctor J by the single adjacency matrix.

$$J = \begin{bmatrix} 0 & M^* & K^* \\ M & 0 & L^* \\ K & L & 0 \end{bmatrix}.$$

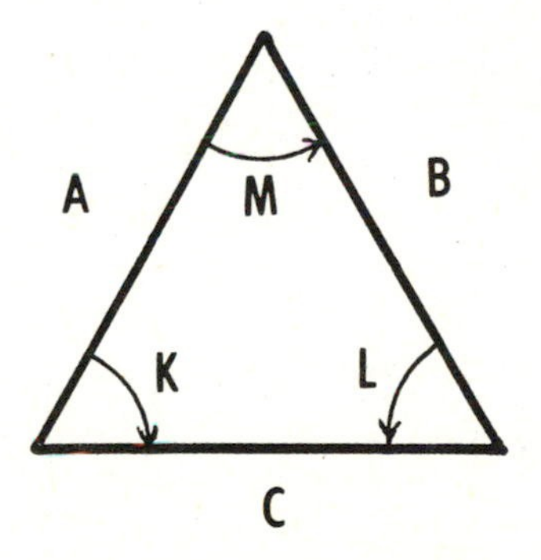

FIG. 26. Symbolic junctor.

We will sometimes use the symbolic representative of the graph and the adjacency matrices shown in Figure 26. Actual junctors are shown in Figures 27 and 28. The adjacency matrices for the

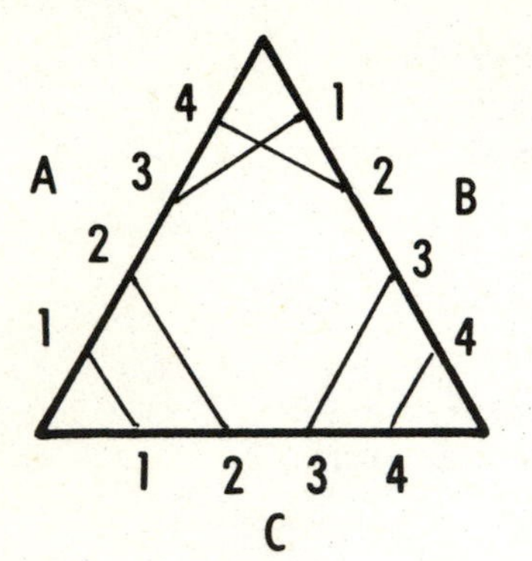

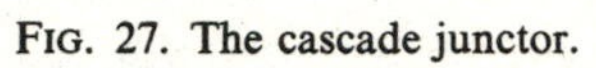
FIG. 27. The cascade junctor.

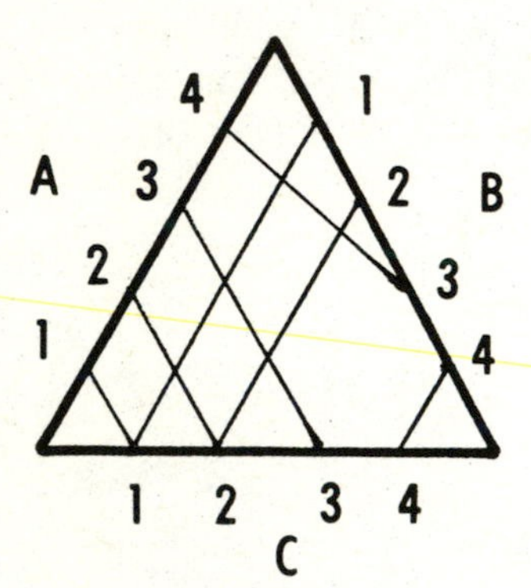

FIG. 28. The hybrid junctor.

hybrid junctor are:

$$K = \begin{bmatrix} 1 & 0 & 0 & 0 \\ 0 & 1 & 0 & 0 \\ 0 & 0 & 1 & 0 \\ 0 & 0 & 0 & 0 \end{bmatrix}, \qquad L = \begin{bmatrix} 1 & 0 & 0 & 0 \\ 0 & 1 & 0 & 0 \\ 0 & 0 & 0 & 0 \\ 0 & 0 & 0 & 1 \end{bmatrix},$$

$$M = \begin{bmatrix} 0 & 0 & 0 & 0 \\ 0 & 0 & 0 & 0 \\ 0 & 0 & 0 & 1 \\ 0 & 0 & 0 & 0 \end{bmatrix}.$$

A *terminal bank* (or box) is a device with n terminals. No mathematical properties of this device are assumed here; in the application which motivated this work, the terminal bank is an electrical network with n terminals but other interpretations can no doubt be given. Given two terminal banks R, S, and a junctor J, we may form a new terminal bank $J(R, S)$ by identifying the terminals of R with the vertices of class A, the terminals of S with vertices of class B, and calling the vertices of class C the terminals of the new terminal bank.

If we have three terminal banks, R, S, and T, and two copies of the junctor J, we may form the terminal box $J(R, J(S, T))$ with adjacency matrix

$J(R, J(S, T)) =$

	R	S	T	new
R	0	M^*K	M^*L	K^*
S	K^*M	0	M^*	K^*L
T	L^*M	M	0	L^{*2}
new	K	LK	L^2	0

(I)

This connection is shown symbolically in Figure 29.

Alternatively, we may form the terminal bank $J(J(R, S), T)$ with adjacency matrix

$J(J(R, S), T) =$

	R	S	T	new
R	0	M^*	K^*M^*	K^{*2}
S	M	0	L^*M^*	L^*K^*
T	MK	ML	0	L^*
new	K^2	KL	L	0

(II)

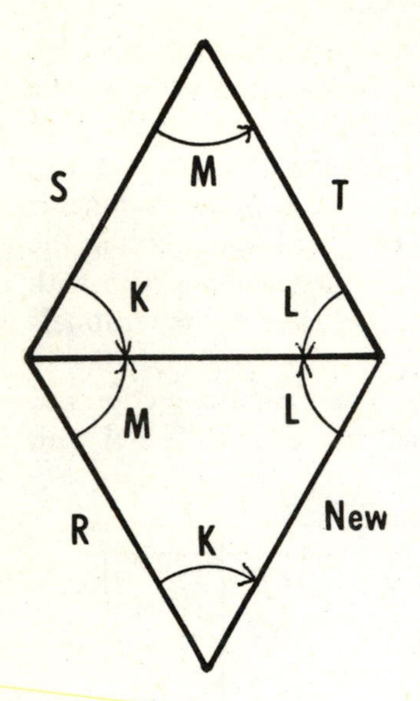

FIG. 29. $J(R, J(S, T))$.

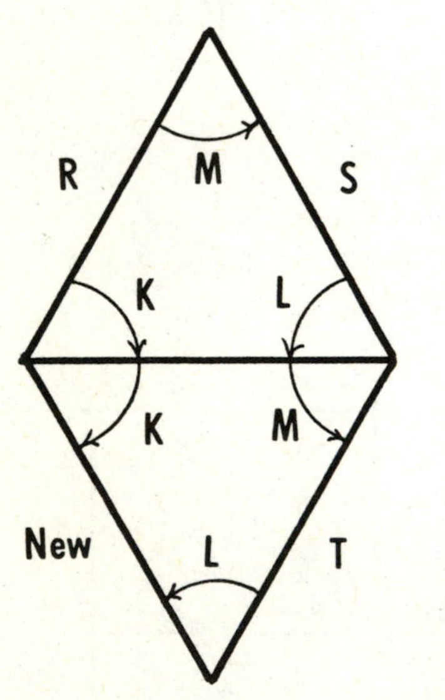

FIG. 30. $J(J(R, S), T)$.

and symbolic representation given in Figure 30. In case $J(R, J(S, T))$ and $J(J(R, S), T)$ define the same connections between the terminals of R, S, T and the new terminals, we say that the junctor is *associative*.

Most of the junctors arising in the study of electrical networks are associative, but the non-associative ones are of some interest. An example of an associative junctor is the hybrid junctor.

We now characterize all associative junctors.

THEOREM A: *The junctor J is associative if and only if the vertices may be renumbered—the same permutation being used in all three classes—so that the matrices K, L, and M are*

$$K = \begin{bmatrix} I & 0 & 0 \\ 0 & I & 0 \\ 0 & 0 & 0 \end{bmatrix}, \qquad L = \begin{bmatrix} I & 0 & 0 \\ 0 & 0 & 0 \\ 0 & 0 & I \end{bmatrix},$$

$$M = \begin{bmatrix} N & 0 & 0 \\ 0 & 0 & P \\ 0 & 0 & 0 \end{bmatrix},$$

where P is a permutation matrix, and N is a diagonal matrix.

Proof: We omit the exact details, but summarize the basic idea. We want to show that two vertices are connected by a path in $J(R, J(S, T))$, if and only if they are connected by a path in $J(J(R, S), T)$. Since the product of adjacency matrices specifies paths in a graph and we have the adjacency matrix representations of (I) and (II), we need only compute the various connections.

Theorem A can be used to show that the series, the parallel, the hybrid and the cascade junctors are all associative.

REFERENCES

1. Ahlfors, L. V., and L. Sario, *Riemann Surfaces*. Princeton: Princeton University Press, 1960, 214–228. MR **22** #5729.

2. Anderson, W. N. Jr., and R. J. Duffin, "Series and parallel addition of matrices," *J. Math. Anal. Appl.*, **26** (1969), 576–594. MR **39** #3904.

3. Birkhoff, G., "Lattice theory," 3rd ed., *Amer. Math. Soc. Colloq. Publ.*, **25** (1967). MR **37** #2638.

4. Bott, R., and R. J. Duffin, "On the algebra of networks," *Trans. Amer. Math. Soc.*, **74** (1953), 99–109. MR **15** #95.

5. ——, "On the Wang algebra of networks," *Bull. Amer. Math. Soc.*, **57** (1951), 136.

6. Brooks, R. L., C. A. B. Smith, A. H. Stone and W. T. Tutte, "The dissection of rectangles into squares," *Duke Math. J.*, **7** (1940), 312–340. MR **2** #153.

7. Derrick, W. R., "A weighted volume-diameter inequality for *N*-cubes," *J. Math. Mech.*, **18** (1968/69), 453–472. MR **39** #7508.

8. Duffin, R. J., "Impossible behavior of nonlinear networks," *J. Appl. Phys.*, **26** (1955), 603–605. MR **16** #1077.

9. ——, "Distributed and lumped networks," *J. Math. Mech.*, **8** (1959), 793–826. MR **21** #4766.

10. ——, "An analysis of the Wang algebra of networks," *Trans. Amer. Math. Soc.*, **93** (1959), 114–131. MR **22** #49.

11. ——, "The extremal length of a network," *J. Math. Anal. Appl.*, **5** (1962), 200–215. MR **26** #1024.

12. ——, "Topology of series-parallel networks," *J. Math. Anal. Appl.*, **10** (1965), 303–318. MR **31** #85.

13. ——, "Estimating Dirichlet's integral and electrical resistance for systems which are not self-adjoint," *Arch. Rational Mech. Anal.*, **30** (1968), 90–101. MR **37** #3812.

14. ——, "Potential theory on a rhombic lattice," *J. Combinatorial Theory*, **5** (1968), 258–272. MR **38** #331.

15. ——, "Optimum heat transfer and network programming," *J. Math. Mech.*, **17** (1968), 759–768.

16. Duffin, R. J. and T. A. Porsching, "Bounds for the conductance of a leaky plate via network models," *Proc. Sympos. on Generalized Networks*, Polytechnic Institute of Brooklyn (1966).

17. Erickson, K. E., "A new operation for analyzing series-parallel networks," *IEEE Trans. Circuit Theory*, **CT-6** (1959), 124–126.

18. Lehman, A., "Problem 60-5-A resistor network inequality," *SIAM Rev.*, **4** (1962), 150–155.

19. ——, *The width-length inequality*, unpublished report (1964).

20. Minty, G., "On the axiomatic foundations of the theories of directed linear graphs, electrical networks and network-programming," *J. Math. Mech.*, **15** (1966), 485–520. MR **32** #5543.

21. Lord Rayleigh, "On the approximate solution of certain problems relating to the potential," *Proc. London Math. Soc.*, **7** (1876), 70–75: *Sci. Papers* **1**, 39.

22. ——, "The theory of sound," **2**, 2nd ed., London: Macmillan, 1896, 305–308.

23. Riordan, J., and C. E. Shannon, "The number of two-terminal series-parallel networks," *J. Mathematical Phys.*, **21** (1942), 83–93. MR **4** #151.

24. Weyl, H., "Reparticion de corriente en una red conductora," *Rev. Mat. Hisp.-Amer.*, **5** (1923), 153–164.

25. Duffin, R. J., D. Hazony and N. Morrison, "Network synthesis through hybrid matrices," *SIAM J.*, **14** (1966), 390–413. MR **34** #7275.

26. Anderson, W. N., Jr., R. J. Duffin and G. E. Trapp, "Parallel subtraction of matrices", *Proc. Nat. Acad. Sci.*, **69** (1972), 2530–2531.

27. Anderson, W. N., Jr., R. J. Duffin and G. E. Trapp, "Tripartite graphs to analyze the interconnection of networks", *Graph Theory and Applications*, 1–12. Edited by Y. Alavi, D. R. Lick and A. T. White, New York: Springer-Verlag, 1972.

28. Bhargava, S. and R. J. Duffin, "Network models for maximization of heat transfer under weight constraints", *Networks*, **2** (1972), 285–299.

29. Derrick, W. R., "Extremal length and Rayleigh's reciprocal theorem", *Applicable Analysis* (1973).

30. Duffin, R. J., "Nonlinear networks," IIa, *Bull. Amer. Math. Soc.*, **53** (1947), 963–971.

31. ——, "Dual programs and minimum cost", *SIAM J.*, **10** (1962), 119–123.

32. Duffin, R. J. and D. K. McLain, "Optimum shape of a cooling fin on a convex cylinder", *J. Math. Mech.*, **17** (1968), 769–784.

33. Duffin, R. J., "Duality inequalities of mathematics and science", *Nonlinear Programming*, 1970, 401–423. Edited by J. B. Rosen, O. L. Mangasarian and K. Ritter, New York: Academic Press.

34. Duffin, R. J. and G. E. Trapp, "Hybrid addition of matrices—network theory concept", *Applicable Analysis*, **2** (1972), 241–254.

35. Minty, G. J., "Monotone networks", *Proc. Roy. Soc. London A257*, (1960), 194–212.

36. Mitra, S. K. and C. R. Rao, "Theory and application of a constrained inverse", *SIAM J. Appl. Math.*, **24** (1973), 473–478.

37. Mitra, S. K. and M. L. Puri, "On parallel sum and difference of matrices", *J. Math. Anal. Appl.*, **44** (1973), 92–97.

38. Penfield, P., R. Spence, and S. Duinker, *Tellegen's Theorem and Electrical Networks*, Cambridge: MIT Press, 1970.

39. Rao, C. R. and S. K. Mitra, *Generalized Inverses of Matrices and its Applications*, New York: Wiley, 1971.

40. Hoffman, A. J., and J. G. Kruskal, "Integral boundary points of convex polyhedra", *Ann. of Math. Studies No.* **38** (1956), 223–246, Edited by H. W. Kuhn and A. W. Tucker, Princeton University Press.

41. Whitney, H., "On the abstract properties of linear dependence", *Amer. J. Math.*, **57** (1935), 504–533.

42. Tutte, W. T., "A class of abelian groups", *Canad. J. Math.*, **8** (1956), 13–28.

43. Moore, E. F., and C. E. Shannon, "Reliable circuits using less reliable relays", *J. Franklin Inst.*, **262** (1956), 191–208, 281–298.

FLOW NETWORKS AND COMBINATORIAL OPERATIONS RESEARCH

D. R. Fulkerson

PART I: FLOWS IN NETWORKS

1. Maximal flow. A *directed network* (graph) $G = [N; \mathcal{A}]$ consists of a finite collection N of elements $1, 2, \ldots, n$ together with a subset $\mathcal{A}$ of the ordered pairs (i, j) of distinct elements of N. The elements of N will be called *nodes*; members of $\mathcal{A}$ are *arcs*. Figure 1.1 shows a directed network having four nodes and six arcs (1, 2), (1, 3), (2, 3), (2, 4), (3, 2), and (3, 4).

Sometimes we shall also consider *undirected networks*, for which the set $\mathcal{A}$ consists of unordered pairs of nodes. For emphasis, these will then be termed *links*.

Suppose that each arc (i, j) of a directed network has associated with it a nonnegative number c_{ij}, the *capacity* of (i, j), to be thought of as representing the maximal amount of some commodity that can arrive at j from i along (i, j) per unit time in a steady-state situation. Then a natural question is: What is the

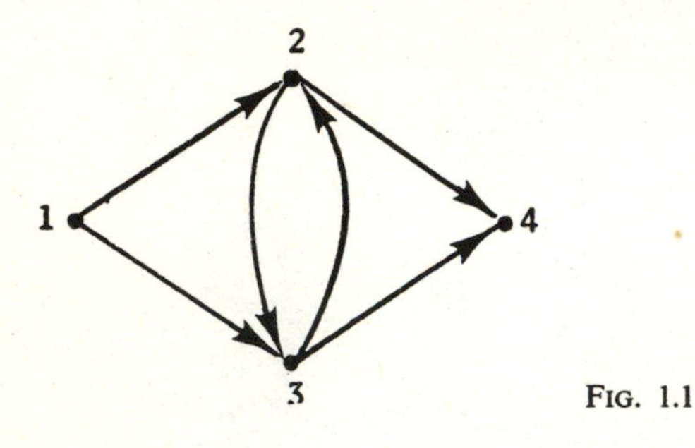

FIG. 1.1

maximal amount of commodity-flow from some node to another via the entire network? (For example, one might think of a network of city streets, the commodity being cars, and ask for a maximal traffic flow from some point to another.) We may formulate the question mathematically as follows: Let 1 and n be the two nodes in question. A *flow, of amount* v, from 1 *to* n in $G = [N; \mathcal{A}]$ is a function x from $\mathcal{A}$ to real numbers (a vector x having components x_{ij} for (i, j) in $\mathcal{A}$) that satisfies the linear equations and inequalities

$$\sum_j x_{ij} - \sum_j x_{ji} = \begin{cases} v, i = 1, \\ -v, i = n, \\ 0, \text{otherwise}, \end{cases} \tag{1.1}$$

$$0 \leq x_{ij} \leq c_{ij}, \qquad (i, j) \text{ in } \mathcal{A}. \tag{1.2}$$

In (1.1), the sums are of course over those nodes for which x is defined. We call 1 the *source*, n the *sink*. A *maximal flow* from source to sink is one that maximizes the variable v subject to (1.1), (1.2).

Figure 1.2 shows a flow from source node 1 to sink node 6 of amount 7. In Figure 1.2, the first number of each pair beside an arc is the arc capacity, the second number the arc flow.

To state the fundamental theorem about maximal flow, we need one other notion, that of a cut. A *cut separating* 1 and n is a partition of the nodes into two complementary sets, I and J, with

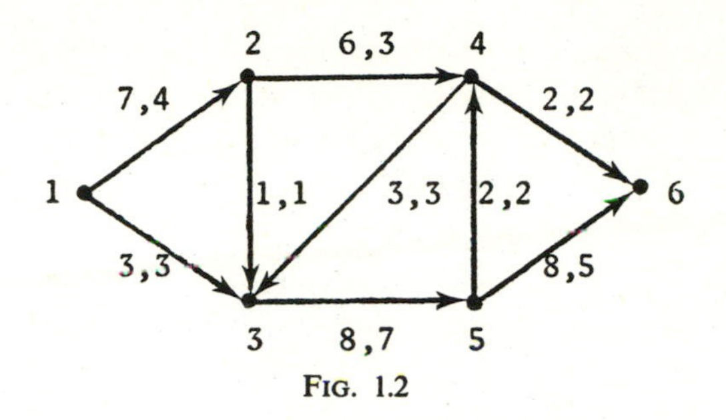

FIG. 1.2

1 in I, say, and n in J. The *capacity* of the cut is then

$$\sum_{\substack{i \text{ in } I \\ j \text{ in } J}} c_{ij}. \tag{1.3}$$

(For instance, if $I = \{1, 3, 4\}$ in Fig. 1.2, the cut has capacity $c_{12} + c_{35} + c_{46} = 17$.) A cut separating source and sink of minimum capacity is a *minimal* cut, relative to the given source and sink.

Summing the equations (1.1) over i in the source-set I of a cut and using (1.2) shows that

$$v = \sum_{\substack{i \text{ in } I \\ j \text{ in } J}} (x_{ij} - x_{ji}) \leqslant \sum_{\substack{i \text{ in } I \\ j \text{ in } J}} c_{ij}. \tag{1.4}$$

In words, for an arbitrary flow and arbitrary cut, the net flow across the cut is the flow amount v, which is consequently bounded above by the cut capacity. Theorem 1.1 below asserts that equality holds in (1.4) for some flow and some cut, and hence the flow is maximal, the cut minimal [11].

THEOREM 1.1: *For any network, the maximal flow amount from source to sink is equal to the minimal cut capacity relative to the source and sink.*

Theorem 1.1 is a kind of combinatorial counterpart, for the special case of the maximal flow problem, of the duality theorem for linear programs, and can be deduced from it [5]. But the most revealing proof of Theorem 1.1 uses a simple "marking" or "labeling" process [12] for constructing a maximal flow, which also yields the following theorem.

THEOREM 1.2: *A flow x from source to sink is maximal if and only if there is no flow-augmenting path with respect to x.*

Here we need to say what an x-augmenting path is. First of all, a path from one node to another is a sequence of distinct end-to-end arcs that starts at the first node and terminates at the second; arcs traversed with their direction in going along the path are *forward* arcs of the path, while arcs traversed against their direction are *reverse* arcs of the path. A path from source to sink is x-augmenting provided that $x < c$ on forward arcs and $x > 0$ on reverse arcs. For example, the path (1, 2), (2, 4), (5, 4), (5, 6) in Figure 1.2 is an augmenting path for the flow shown there. Figure 1.3 indicates how such a path can be used to increase the amount of flow from source to sink.

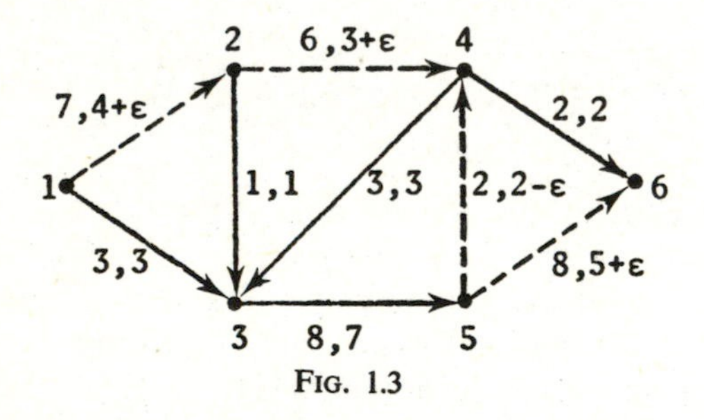

FIG. 1.3

Taking the flow change ϵ along the path as large as possible in Figure 1.3, namely $\epsilon = 2$, produces a maximal flow, since the cut $I = \{1, 2, 4\}$, $J = \{3, 5, 6\}$ is then "saturated." (See Figure 1.4.)

The labeling process of [12] is a systematic and efficient search, fanning out from the source, for a flow-augmenting path. If none such exists, the process ends by locating a minimal cut.

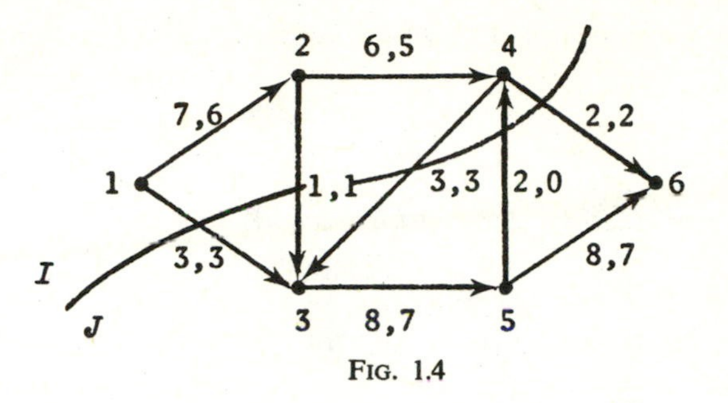

FIG. 1.4

The following theorem, of special significance for combinatorial applications, is also a consequence of the procedure sketched above for constructing maximal flow.

THEOREM 1.3: *If all arc capacities are integers, there is an integral maximal flow.*

It is sometimes convenient to alter the constraints (1.2) of the maximal flow problem to

$$l_{ij} \leqslant x_{ij} \leqslant c_{ij}. \tag{1.5}$$

Here l is a given lower bound function satisfying $l \leqslant c$. The analogue of Theorem 1.1 is then

THEOREM 1.4: *If there is a function x satisfying* (1.1) *and* (1.5) *for some number v, then the maximum v subject to these constraints is equal to the minimum of*

$$\sum_{\substack{i \text{ in } I \\ j \text{ in } J}} (c_{ij} - l_{ji}) \tag{1.6}$$

taken over all cuts I, J separating source and sink. On the other *hand, the minimum v is equal to the maximum of*

$$\sum_{\substack{i \text{ in } I \\ j \text{ in } J}} (l_{ij} - c_{ji}) \tag{1.7}$$

taken over all cuts I, J separating source and sink.

The question of the existence of such a flow x, together with another flow feasibility question, will be discussed in the next section.

2. Feasibility theorems. The constraints of the maximal flow problem are, of course, always feasible, since $x = 0$ satisfies (1.1), (1.2) with $v = 0$. By changing the constraints in various ways, interesting feasibility questions arise. Here we shall consider two such, one involving supplies and demands at nodes, the other lower bounds on arc flows, as in (1.5).

Let $G = [N; \mathcal{A}]$ have capacity function c, and let S and T be disjoint subsets of N. With each i in S, associate a *supply* $a_i \geqslant 0$; with each i in T, a *demand* $b_i \geqslant 0$, and impose the constraints

$$\sum_j (x_{ij} - x_{ji}) \begin{cases} \leqslant a_i, & i \text{ in } S, \\ \leqslant -b_i, & i \text{ in } T, \\ = 0, & \text{otherwise}, \end{cases} \tag{2.1}$$

$$0 \leqslant x_{ij} \leqslant c_{ij}, \qquad (i, j) \text{ in } \mathcal{A}. \tag{2.2}$$

In words, the net flow out of i in S is bounded above by the supply a_i, and the net flow into i in T is bounded below by the demand b_i. When are the supply-demand constraints (2.1), (2.2) feasible?

This question is easily answered by applying Theorem 1.1 to an enlarged network. Extend $G = [N; \mathcal{A}]$ to $G^* = [N^*; \mathcal{A}^*]$ by adjoining a source 0 and sink $n + 1$, together with source arcs $(0, j)$ for j in S, and sink arcs $(i, n + 1)$ for i in T. (See Figure 2.1.) The capacity function c^* on $\mathcal{A}^*$ is defined by $c^*_{0,j} = a_j$ for j in S, $c^*_{i,n+1} = b_i$ for i in T, $c^*_{ij} = c_{ij}$ for (i, j) in $\mathcal{A}$. The constraints (2.1)

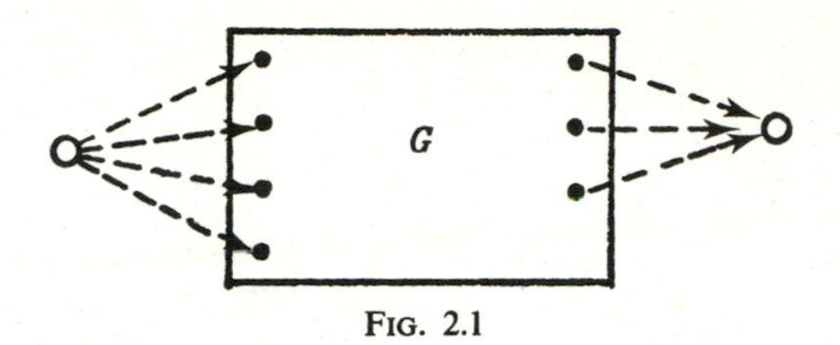

FIG. 2.1

and (2.2) are feasible if and only if the maximal flow amount from source to sink in the enlarged network is at least $\Sigma_{i \text{ in } T} b_i$, that is, if and only if a maximal flow saturates all sink arcs. Hence we need only construct a maximal flow in order to check the feasibility of (2.1), (2.2). By pushing the analysis a little further, using Theorem 1.1, the following theorem emerges [21].

THEOREM 2.1: *The supply-demand constraints* (2.1), (2.2) *are feasible if and only if, for each subset T' of T, there is a flow $x(T')$ that satisfies the aggregate demand $\Sigma_{i \text{ in } T'} b_i$ without violating the supply limitations at nodes of S.*

Here, satisfying the aggregate demand over T' means that the net flow into the set T' must be at least $\Sigma_{i \text{ in } T'} b_i$, without regard for the individual demands in T'. The necessity of the condition is of course clear; sufficiency asserts the existence of a single flow x meeting all individual demands, provided the flows $x(T')$ exist for all subsets T' of T.

It should be noted that if the functions a, b, c of (2.1), (2.2) are integral valued, and if feasible flows exist, then there is an integral feasible flow. This follows from Theorem 1.3 and the conversion of (2.1), (2.2) to a maximal flow problem. A similar integrity statement holds for the situation of Theorem 1.4, and indeed, for all the flow problems to be discussed in any detail in this survey.

We turn now to a consideration of lower bounds on arc flows, as in (1.5), and pose the resulting feasibility question in terms of circulations, i.e., flows that are source-sink free, instead of flows from source to sink. (One can always add a "return-flow" arc from sink to source to convert to circulations.) Thus we are questioning the feasibility of the constraints

$$\sum_j (x_{ij} - x_{ji}) = 0, \qquad i \text{ in } N, \tag{2.3}$$

$$l_{ij} \leqslant x_{ij} \leqslant c_{ij}, \qquad (i,j) \text{ in } \mathcal{A}. \tag{2.4}$$

The following theorem answers the question [26]. Its proof can be made to rely on Theorem 1.1 [15].

THEOREM 2.2: *The constraints* (2.3), (2.4) *are feasible if and only if*

$$\sum_{\substack{i \text{ in } I \\ j \text{ in } J}} c_{ij} \geqslant \sum_{\substack{i \text{ in } I \\ j \text{ in } J}} l_{ji} \tag{2.5}$$

holds for all partitions I, J of N.

Again the necessity is clear, since (2.5) simply says there must be sufficient escape capacity from the set I to take care of the flow forced into I by the function l. But sufficiency is not obvious.

Other useful flow feasibility theorems have been deduced [19, 26]. In each case, Theorem 1.1 can be used as the main tool in a proof.

3. Minimal cost flows. One of the most practical problem areas involving network flows is that of constructing flows satisfying constraints of various kinds and minimizing cost. The standard linear programming transportation problem, which has an extensive literature, is in this category.

We put the problem as follows. Each arc (i,j) of a network $G = [N; \mathcal{A}]$ has a capacity c_{ij} and a cost a_{ij}. It is desired to construct a flow x from source to sink of specified amount v that minimizes the total flow cost

$$\sum_{(i,j) \text{ in } \mathcal{A}} a_{ij} x_{ij} \tag{3.1}$$

over all flows that send v units from source to sink. In many applications one has supplies of a commodity at certain points in a transportation network, demands at others, and the objective is to satisfy the demands from the supplies at minimum cost.

By treating v as a parameter, the method for constructing maximal flows can be used to construct minimal cost flows throughout the feasible range of v. Indeed, the solution procedure can be viewed as one of solving a sequence of maximal flow problems, each on a subnetwork of the original one [14]. Another, not essentially different, viewpoint is provided by the following theorem [1, 29].

THEOREM 3.1: *Let x be a minimal cost flow from source to sink of amount v. Then the flow obtained from x by adding $\epsilon > 0$ to the flow in forward arcs of a minimal cost x-augmenting path, and subtracting ϵ from the flow in reverse arcs of this path, is a minimal cost flow of amount $v + \epsilon$.*

Here the cost of a path is the sum of arc costs over forward arcs minus the corresponding sum over reverse arcs, i.e., the cost of "sending an additional unit" via the path.

Thus, if all arc costs a_{ij} are nonnegative, for example, one can start with the zero flow and apply Theorem 3.1 to obtain minimal cost flows for increasing v. (The cost profile thereby generated is piecewise linear and convex.) All that is needed to make this an explicit algorithm is a method of searching for a minimal cost flow augmenting path. Various ways of doing this can be described. One such will be given in Part II, section 1.

Another method [17] for constructing minimal cost flows poses the problem in circulation format, that is, (3.1) is to be minimized subject to (2.3), (2.4). This construction has a number of advantages, principally in terms of generality and flexibility. For instance, it may be started with any circulation; even (2.4) need not be satisfied initially. Also, no assumption about the cost function is required.

These methods produce integral flows in case the arc capacities (and lower bounds) are integers.

4. Maximal dynamic flow. Suppose that each arc (i, j) of a network G has not only a capacity, but a transit time t_{ij} as well, and that we are interested in determining the maximum amount of flow that can reach sink n from source 1 in a specified number t of time periods. This dynamic flow problem can always be treated as a static flow problem in a time-expanded version G_t of G. For example, if the given network G is that of Figure 4.1 and if each arc of G has unit transit time, then G_3 is shown in Figure 4.2. (We have included "storage arcs" leading from a location to itself one unit of time later.)

Expanding the network in this way puts one back in the static case. Moreover, arc capacities and transit times can vary with time and this is still so. However, if each capacity and transit time is fixed over time, the problem can be solved in the smaller network G. Specifically, a maximal dynamic flow for t periods can be generated from a static flow x in G of amount v that minimizes

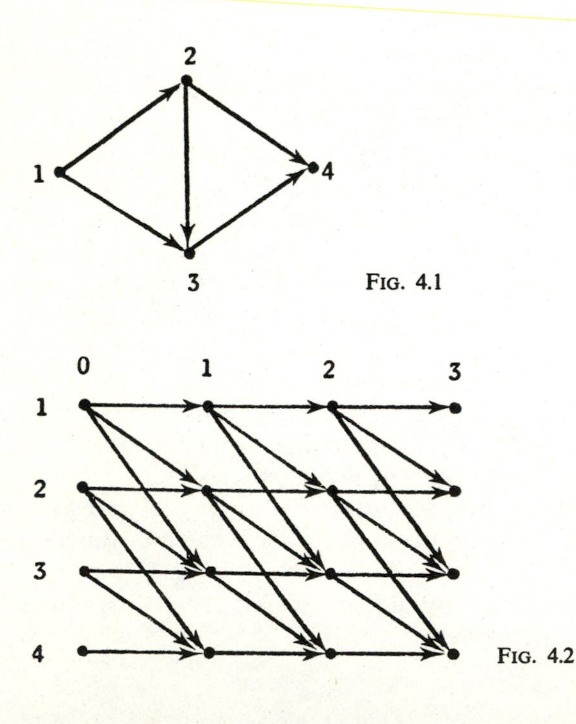

FIG. 4.1

FIG. 4.2

the linear form

$$\sum_{(i,j) \text{ in } \mathcal{A}} t_{ij}x_{ij} - (t + 1)v \tag{4.1}$$

over all flows in G from source 1 to sink n [14]. By adding the return-flow arc $(n, 1)$ to G with $c_{n1} = \infty$, $t_{n1} = -(t + 1)$, the problem may be viewed as one of constructing a circulation that minimizes the "cost" form (4.1).

5. Multi-terminal maximal flow.* Heretofore we have phrased statements in terms of directed networks. In this section we confine the discussion to undirected flow networks, by which we merely mean the following. A link (i, j) can carry flow in either direction and has the same flow capacity each way. Thus one can think of an arc (i, j) with capacity c_{ij} and an arc (j, i) with capacity $c_{ji} = c_{ij}$. The assumption of a symmetric capacity function c makes the results described in this section considerably simpler and more appealing than they would otherwise be.

Instead of dealing with a single source and sink, we shift attention to all pairs of distinct nodes taken as terminals for flows. These flows are not to be thought of as occurring simultaneously.

Let v_{ij} denote the maximal flow amount from i to j. Thus the function v is symmetric, $v_{ij} = v_{ji}$, and may be determined explicitly for an n-node network by solving $n(n - 1)/2$ maximal flow problems. There is, however, a much simpler way of determining the function v, one that involves the solution of only $n - 1$ maximal flow problems; in addition, there is a simple condition in order that a symmetric function v be realizable as the maximal multiterminal flow function of some undirected network [22].

THEOREM 5.1: *A symmetric, nonnegative function v is realizable by an undirected network if and only if v satisfies*

$$v_{ij} \geqslant \min(v_{ik}, v_{kj}) \tag{5.1}$$

for all triples i, j, k.

---

*For a full discussion of this topic, see the paper by Gomory and Hu in this volume.

The necessity of the "triangle inequaltiy" (5.1) follows easily from Theorem 1.1.

The condition (5.1) imposes severe limitations on the function v. For instance, among the three functional values appearing in (5.1), two must be equal and the third no smaller than their common value. It also follows that if the network has n nodes, v can take on at most $n - 1$ numerically different functional values. It is not altogether surprising, therefore, that v can be determined by a simpler process than solving all single-terminal maximal flow problems. This process systematically picks out precisely $n - 1$ cuts in the network having the property that v_{ij} is determined by the minimum one of these cuts separating i and j [22]. For example, in the network of Figure 5.1 the relevant cuts are those shown. Thus, for instance, since nodes 1 and 4 are separated by the three cuts (1/2, 3, 4, 5, 6), (1, 3/2, 4, 5, 6), (1, 2, 3, 5/4, 6) having capacities 8, 6, 6 respectively, then $v_{14} = \min(8, 6, 6) = 6$.

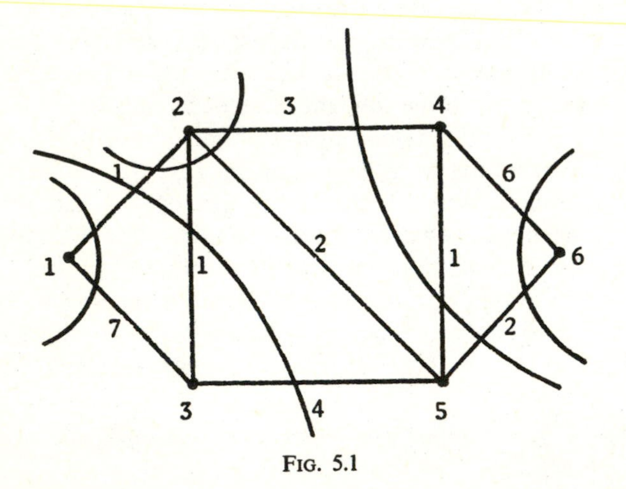

FIG. 5.1

6. Other flow problems. The flow problems that have been discussed thus far all have the useful and pleasant feature that the assumption of integral data implies the existence of an integral solution. A number of flow problems that do not share this property have also been studied. Among these, we mention flows in networks with gains [29], simultaneous multi-terminal flows [13], and problems involving optimal synthesis of flow networks that meet specified requirements [16, 22]. Methods of solution for such problems are, in general, more complicated than methods for those we have discussed. One rather surprising exception to this statement is the following synthesis problem: Suppose it is desired to construct an undirected network on a specified number of nodes so that $v_{ij} \geqslant r_{ij}$ for stipulated requirements r_{ij}, with the total sum of link capacities of the network minimal. A very simple combinatorial method of solution for this synthesis problem is given in [22].

PART II: COMBINATORIAL PROBLEMS

1. Network potentials and shortest chains. Consider a directed network in which each arc (i, j) has associated with it a positive number a_{ij}, which may be thought of as the length of the arc, or the cost of traversing the arc. How does one determine a shortest chain from some node to another? Here we have used *chain* to mean a path containing only forward arcs, the length of the chain being obtained by adding its arc lengths.

While this is a purely combinatorial problem, it may also be viewed as a flow problem simply by imposing a cost a_{ij} per unit flow in (i, j), taking all arc capacities infinite, and asking for a minimal cost flow of one unit from the first node to the second. An integral optimal flow corresponding to $v = 1$ singles out a shortest chain.

Many ways of locating shortest chains efficiently have been suggested. We describe one [10]. Like others, it simultaneously finds shortest chains from the first node to all others reachable by chains.

In this method, each node i will initially be assigned a number π_i. These node numbers, which we shall refer to as *potentials*, will then be revised in an iterative fashion. Let 1 be the first node. To start, take $\pi_1 = 0$, $\pi_i = \infty$ for $i \neq 1$. Then search the list of arcs for an arc (i, j) whose end potentials satisfy

$$\pi_i + a_{ij} < \pi_j. \tag{1.1}$$

(Here $\infty + a = \infty$.) If such an arc is found, change π_j to $\pi_j' = \pi_i + a_{ij}$, and search again for an arc satisfying (1.1), using the new node potentials. Stop the process when the node potentials satisfy

$$\pi_i + a_{ij} \geqslant \pi_j \tag{1.2}$$

for all arcs.

It is not hard to show that the process terminates, and that when this happens, the potential π_j is the length of a shortest chain from 1 to j. (Here $\pi_j = \infty$ at termination means there is no chain from 1 to j.) A shortest chain from 1 to j can be found by tracing back from j to 1 along arcs satisfying (1.2) with equality (see Figure 1.1).

Practical applications that require shortest chains are numerous. For instance, in making up a table of highway distances between cities, a shortest chain between each pair needs to be found. A less

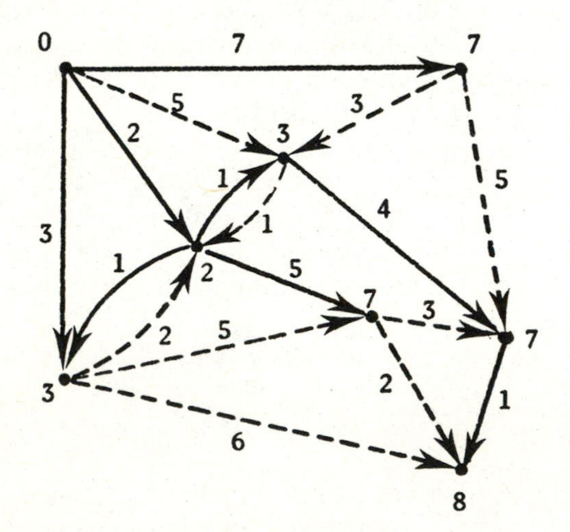

FIG. 1.1

obvious application is the discrete version of the problem of determining the least time for an airplane to climb to a given altitude [2]. Some other applications will be discussed in following sections.

While we have assumed positive lengths for the method described above, this assumption can be weakened. Call a chain of arcs leading from a node to itself a *directed cycle*. Then it is enough to suppose that all directed cycle lengths are nonnegative.

If directed cycle costs are nonnegative, the minimum cost flow problem of Part I, section 3, can be solved by repeatedly finding cheapest chains in suitable networks. Because of the assumption on the cost function a, we may start with the zero flow. Thus, using Theorem 3.1, it is enough to reduce the problem of finding a cheapest flow-augmenting path with respect to a minimal cost flow x of amount v to that of finding a cheapest chain. Define a new network $G' = [N; \mathcal{A}']$ from the given one $G = [N; \mathcal{A}]$ and the flow x as follows: First, note that we may assume $x_{ij} \cdot x_{ji} = 0$, since $a_{ij} + a_{ji} \geqslant 0$. Now put (i, j) in $\mathcal{A}'$ if either $x_{ij} < c_{ij}$ or $x_{ji} > 0$ and define a' by

$$a'_{ij} = \begin{cases} a_{ij} & \text{if} \quad x_{ij} < c_{ij} \quad \text{and} \quad x_{ji} = 0, \\ -a_{ji} & \text{if} \quad x_{ji} > 0. \end{cases} \tag{1.3}$$

Thus a chain from source to sink in the new network corresponds to an x-augmenting path in the old, and these have the same cost. Moreover, since x is a minimal cost flow, the function a' satisfies the nonnegative directed cycle condition. Hence the method of this section can be used to construct minimal cost flows of successively larger amounts.

2. Optimal chains in acyclic networks. If the network is acyclic (contains no directed cycles), the shortest chain method of the last section can be modified in such a way that, once a potential is assigned a node, it remains unchanged. One can begin by numbering the nodes so that if (i, j) is an arc, then $i < j$. Such a numbering can be obtained as follows: Since the network is acyclic, there are nodes having only outward-pointing arcs. Number these nodes $1, 2, \ldots, k$ in any order. Next delete these nodes and all their arcs, search the new network for nodes having only outward-pointing arcs, and number these, starting with $k + 1$. Repetition of this process leads to the desired kind of numbering (see Figure 2.1).

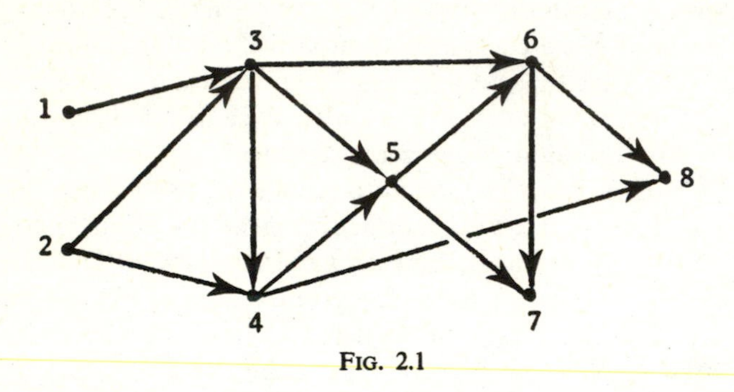

FIG. 2.1

If we wish to find shortest chains from node k to all other nodes reachable from k by chains, the calculation is now trivial. Simply define $\pi_k, \pi_{k+1}, \ldots, \pi_n$ recursively by

$$\begin{cases} \pi_k = 0 \\ \cdots \\ \pi_j = \min\limits_{k \leq i < j} (\pi_i + a_{ij}), \quad j = k + 1, \ldots, n. \end{cases} \tag{2.1}$$

Here the minimum is of course taken over i such that (i, j) is an arc.

Longest chains in acyclic networks can be computed by replacing "min" by "max" in (2.1).

The recursion (2.1), of dynamic programming type, can be applied in a number of problems. We shall discuss three such applications in the following sections.

3. The knapsack problem. Suppose there are K objects, the i-th object having weight w_i and value v_i, and that it is desired to find the most valuable subset of objects whose total weight does not exceed W. Thus we wish to maximize

$$\sum_{i \text{ in } S} v_i \tag{3.1}$$

over subsets $S \subseteq \{1, 2, \ldots, K\}$ such that

$$\sum_{i \text{ in } S} w_i \leqslant W. \tag{3.2}$$

We take $w_1, w_2, \ldots, w_K, W$ to be positive integers.

This combinatorial problem, commonly referred to as the knapsack problem, can be viewed as one of finding a longest chain in a suitable acyclic network. Let the network have nodes denoted by ordered pairs (i, w), $i = 0, 1, \ldots, K$, $w = 0, 1, \ldots, W$. The node (i, w) has two arcs leading into it, one from $(i - 1, w)$, the other from $(i - 1, w - w_i)$, provided these exist. (See Figure 3.1.) The length of the first arc is zero; the other has length v_i. In addition we put in a starting node and join it to all of the nodes $(0, w)$ by arcs of length zero. Then chains from the starting node to (i, w) correspond to subsets of the first i objects whose total weight is at most w, the length of the chain being the value of the subset.

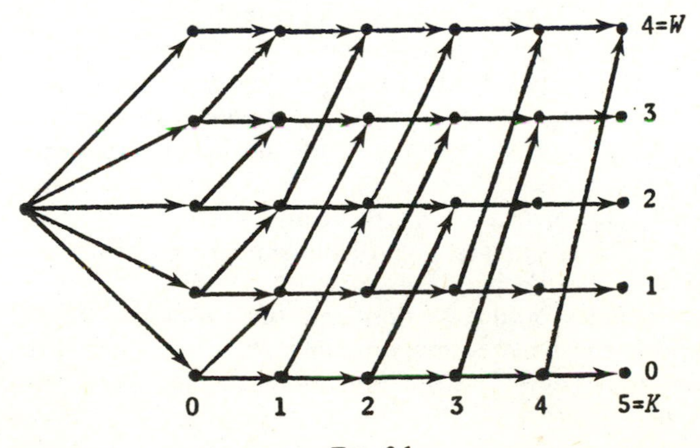

FIG. 3.1

4. Equipment replacement. As equipment deteriorates with age, and improved equipment becomes available on the market, a time may be reached when the purchase cost of new equipment is repaid by its potential future earnings. One is then faced with the problem of determining an optimal replacement policy [9].

For simplicity, consider a single machine and suppose that at the beginning of each of K periods of time it must be decided whether to keep the machine another period or purchase a new one. Let $r(i, t)$ denote the revenue obtainable during period i from a machine which starts the period at age t (the function r may reflect upkeep costs), and let $c(i, t)$ denote the cost of replacing a machine of age t with a new machine if the replacement occurs at the beginning of period i. Thus replacing a machine of age t at period i gives a net return for the period of $r(i, 0) - c(i, t)$.

The acyclic network shown in Figure 4.1 indicates one formulation in terms of chains. Again nodes are points (i, t), $i = 0, 1, \ldots, K$, $t = 0, 1, \ldots, T$. (Here T is some sufficiently large integer; if we start with a machine of age t, $T = K + t$ will do.) In general, two arcs, reflecting the possibilities of keeping or replacing, lead from (i, t), the "keep" arc going to $(i + 1, t + 1)$, the "replace" arc to $(i + 1, 1)$. The first of these has length $r(i, t)$, the second has length $r(i, 0) - c(i, t)$. We may also put in a sink node with arcs of length zero leading into it from the nodes (K, t), $t = 0, 1, \ldots, T$. Then chains from $(0, t)$ to the sink correspond to the various replacement policies starting with a machine of age t, the length of the chain being the total return from the policy.

A simpler network for this problem is shown in Figure 4.2. In Figure 4.2 an arc (i, j) corresponds to keeping the machine throughout periods $i, i + 1, \ldots, j - 1$ and replacing it at the start of period j, the associated length being the return obtained from this action. Thus a longest chain from source 1 to sink K is to be found.

The examples of this section and the preceding section are typical discrete dynamic programming problems. Such problems can always be viewed as seeking optimal chains in appropriate acyclic networks.

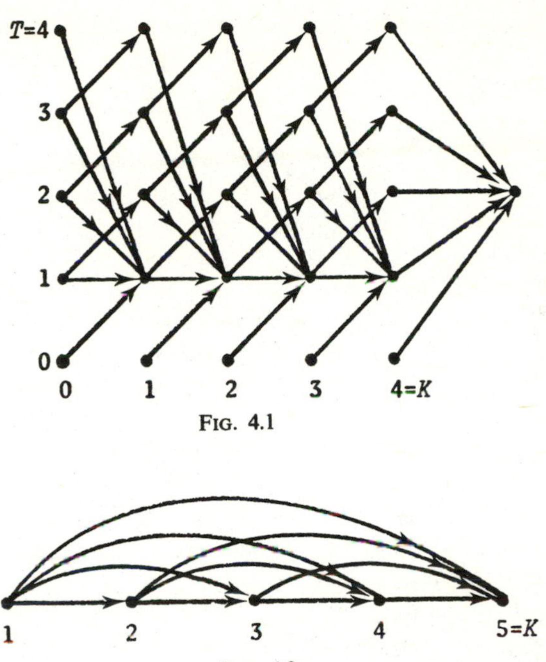

FIG. 4.1

FIG. 4.2

5. Project planning. One of the most popular combinatorial applications involving networks deals with the planning and scheduling of large, complicated projects [33]. Suppose that a project of some kind (the construction of a bridge, for example) is broken down into many individual jobs. Certain of these jobs will have to be finished before others can be started. We may depict the order relations among the jobs by means of an acyclic network whose arcs represent jobs. To take a simple case, suppose there are five jobs with the ordering: 1 precedes 3; 1 and 2 precede 4; 1, 2, 3 and 4 precede 5. This may be pictured by the network shown in Figure 5.1.

Notice that we have added a "dummy" job, the dotted arc of Figure 5.1, to maintain the proper order relations among the jobs.

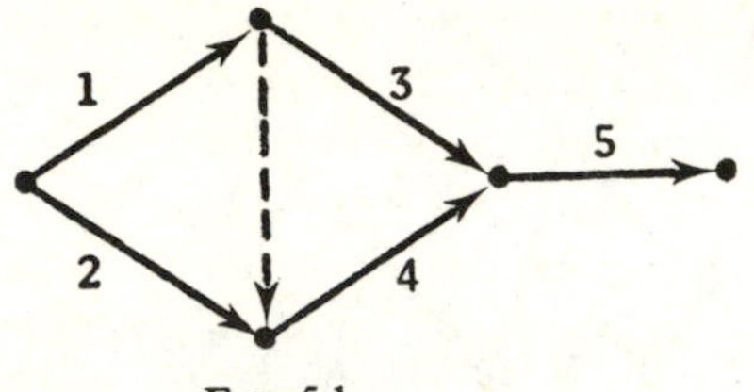

FIG. 5.1

The use of dummies permits a network representation of this kind for any project (finite partially ordered set).

Assuming that each job has a known duration time (dummies have zero duration time), that the only scheduling restriction is that all inward-pointing jobs at a node must be finished before any outward-pointing job can be started, it follows that the minimum time to complete the entire project is equal to the length of a longest chain of jobs. Hence the minimum project duration time can be calculated easily.

Although a fixed time has been assumed for each job, it may be the case that by spending more money, a job can be expedited. The question then arises: Which jobs should money be spent on and how much, in order that the project be finished by a given date at minimum cost? If the time-cost relation for each job is linear, this problem can be shown to be a minimal cost flow problem of the kind described in Part I, section 3 [18, 33].

6. Minimal chain coverings of acyclic networks. The following question concerning acyclic networks has both theoretical and practical interest: What is the minimum number of chains required to cover a given subset of arcs? We first show how flows may be used to answer this question, and then give a practical interpretation.

Let the subset of arcs be denoted by $\mathcal{A}'$. To rephrase the question, we seek the minimum number of chains in the acyclic network G such that every arc of $\mathcal{A}'$ belongs to at least one of the chains. Theorems 1.3 and 1.4 can be used to provide an answer to the question; as sources in G take all nodes with only outward-pointing arcs; as sinks take all nodes with only inward-pointing arcs. Now place a lower bound of 1 on flow in arcs of $\mathcal{Q}'$, 0 on arcs not in $\mathcal{Q}'$, and take all arc capacities infinite. Then an integral flow through G of amount v picks out v chains in G that cover all arcs of $\mathcal{Q}'$, and the second half of Theorem 1.4 implies

THEOREM 5.1: *The minimum number of chains in an acyclic network needed to cover a subset of arcs is equal to the maximum number of arcs of the subset having the property that no two belong to any chain.*

Theorem 5.1 is a mild generalization of a known result on chain decompositions of partially ordered sets [8].

A practical instance of this situation arises if we think of an airline, say, attempting to meet a fixed flight schedule with the minimum number of planes, all of the same type [4]. Let the individual flights be numbered 1, 2, . . . , n. Start and finish times $s_i < f_i$ are known for each flight, and the times t_{ij} to return from the destination of the ith flight to the origin of the jth flight are also known. The flights can be partially ordered by saying that i precedes j if $f_i + t_{ij} \leqslant s_j$, and the resulting partially-ordered set represented by an acyclic network (as in the preceding section). A chain in the network represents a possible assignment of flights to one aircraft. The problem then is to cover the nondummy arcs (those corresponding to actual flights) with the minimum number of chains. Theorem 5.1 asserts that this number is equal to the maximum number of flights, no two of which can be accomplished by a single plane.

Problems of this nature become considerably more complicated if the assumption of a fixed schedule is dropped. For instance, suppose the times s_i, f_i are at our disposal subject to the restriction that $f_i - s_i = t_i$, with the duration times t_i known, as well as the reassignment times t_{ij}. The problem might then be to arrange a schedule completing all flights by a given time and requiring the minimum number of planes, or to finish all flights at the earliest possible time with a fixed number of planes. For such scheduling problems, there is very little known in the way of general theoretical results or good computational procedures. However, some special results have been deduced [27, 31].

7. Assignment problems. The following is typical of an important and well-known class of combinatorial problems having network flow formulations. Suppose there are m men and n jobs, and that it is known whether or not man i is qualified to fill job j, $i = 1, 2, \ldots, m, j = 1, 2, \ldots, n$. When is it possible to fill all jobs with qualified men and how does one determine such an assignment?

Using Theorem 1.3, the problem may be phrased in terms of flows. Corresponding to man i take a source node i, to job j a sink node j, and direct an arc from i to j if man i is qualified for job j. (See Figure 7.1.) Impose a demand of 1 unit at each sink and let each source have a supply of 1 unit. All arc capacities may be taken infinite. The problem of assigning men to jobs thus becomes that of constructing a flow (integral, of course) meeting the demands from the supplies.

Combinatorial interpretations of Theorems 1.1 and 2.1 for this situation lead in the first instance to a well-known theorem about maximum matchings and minimum covers in bipartite networks [35], and in the second instance to an equally well-known, and

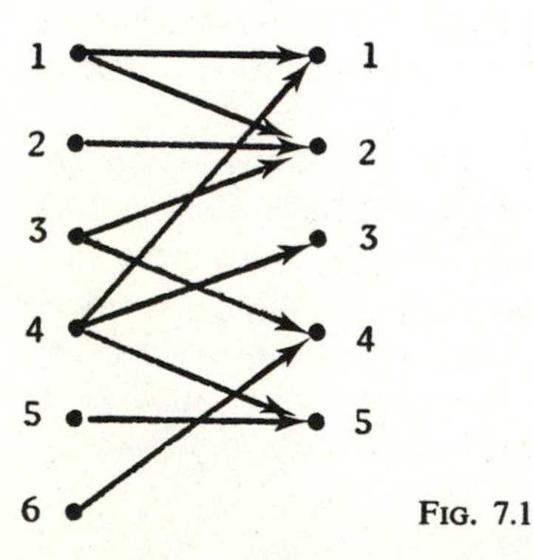

FIG. 7.1

equivalent, theorem concerning systems of distinct representatives for subsets of a given set [24].

A more general assignment problem, usually referred to as that of optimal assignment [36, 42], assumes that man i in job j is worth a_{ij} units, and the total worth of an assignment is given by the sum of the numbers a_{ij} taken over the individual man-job matchings in the assignment. The problem then is to construct an assignment of maximal worth. By taking the cost per unit of flow in arc (i, j) to be $-a_{ij}$, the optimal assignment problem is seen to be a special case of the minimal cost flow problem.

More complicated personnel-assignment models have been formulated in terms of flow networks. For instance, one which involves the recruiting, training, and retraining of personnel to meet stipulated requirements in various job specialties over time has been treated in this way [23].

Applications of the optimal assignment model to other kinds of problems are very numerous. We mention one which involves the optimal depletion of inventory [7]. Suppose a stockpile consists of m items of the same kind, and that the age t_i of item i is known. Also known is a function $u(t)$ giving the utility for an item of age t when withdrawn from the stockpile, together with a schedule of demands specifying the times at which items will be required. The problem is to determine that order of item issue which maximizes the total utility while meeting the demand schedule. (For a concrete example, suppose one has m bottles of wine in his cellar, the ages of each being known, and consumes one bottle of wine weekly. The utility function for wine might appear as in Figure 7.2.) The utility of item i issued at time j is given by $u_{ij} = u(t_i + j)$, and hence the problem is to find an assignment of items to times which is optimal in terms of the u_{ij}.

If the utility function is convex, there is a simple rule for solving the problem: Issue the youngest item first, then the next youngest, and so on. This policy, sometimes called LIFO (last in first out), may be shown optimal here by a simple interchange argument. Similarly, if the utility function is concave, the reverse rule FIFO (first in first out) solves the problem. In general, however, no such simple rule works and an optimal assignment needs to be computed.

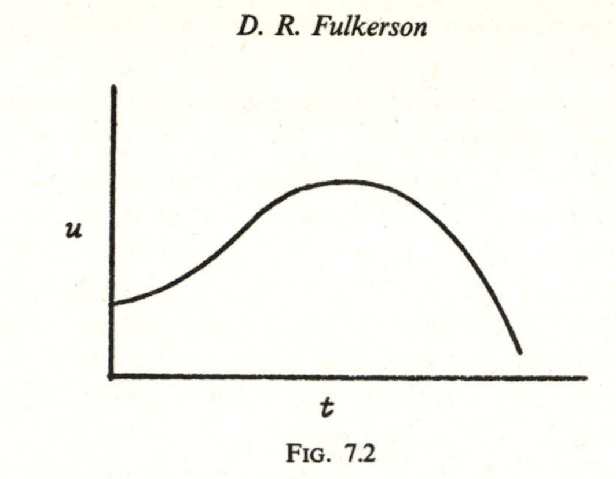

FIG. 7.2

8. Production and inventory planning. Problems involving dynamic production and inventory programs for a single type item have received considerable study. A very simple deterministic problem in this category is the following: Suppose there are n periods of time with known period demands $b_1, b_2, \ldots, b_n$ for the item, that the unit cost of production in period i is p_i, and the unit cost of storage from period i to $i + 1$ is s_i. What pattern of production and storage meets the demands at minimum cost?

The network shown in Figure 8.1 assumes that production in period i can be used to satisfy demand in period i. The ith "production" arc (source arc) has infinite capacity and cost p_i; the ith "storage" arc has infinite capacity and cost s_i; the ith "demand" arc (sink arc) has capacity b_i and zero cost. The problem then is to determine a flow of amount $v = \Sigma_i b_i$ from source to sink that minimizes cost. Clearly, production and storage capacities may be introduced if desired. But if these are left infinite, there is a very simple rule for solving the problem. For the ith demand, compare the chain costs

$$\begin{cases} p_1 + s_1 + \cdots + s_{i-1} \\ p_2 + s_2 + \cdots + s_{i-1} \\ \cdots \\ p_i \end{cases}$$

and take the smallest of these. Then send b_i units of flow along the corresponding chain. An almost equally simple rule works in case each period's production cost is convex in the number of items produced [30].

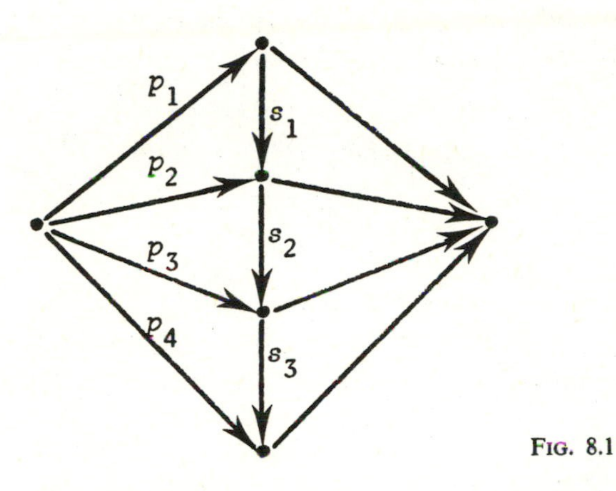

FIG. 8.1

If it is assumed that demands do not have to be satisfied, but that unfulfilled demand in period i results in a penalty cost c_i per unit, we may place a flow cost $-c_i$ on the ith sink arc and solve the minimal cost flow problem parametrically in the flow amount v, selecting that v which gives the least cost.

9. Optimal capacity scheduling. The model of this section, proposed and studied in [41], is a rather general one which can be shown to include several of those previously discussed here. One version of the model is described in [41] as follows: "A decision maker must contract for warehousing capacity over n time periods, the minimal capacity requirement for each period being deterministically specified. His economic problem arises because savings may possibly accrue by his undertaking long-term leasing or contracting at favorable periods of time, even though such commitments may necessitate leaving some of the capacity idle during several periods."

To put the problem mathematically, let d_i be the minimal capacity demand in period i. Let x_{ij}, $i < j$, be the number of units of capacity acquired at the beginning of period i, available for possible use during periods $i, i+1, \ldots, j-1$, and relinquished at the beginning of period j, and let a_{ij} be the associated unit cost. Then the problem is to find $x_{ij} \geqslant 0$ that minimize

$$\sum_{i=1}^{n} \sum_{j=i+1}^{n+1} a_{ij} x_{ij} \tag{9.1}$$

subject to the constraints

$$\sum_{i=1}^{k} \sum_{j=k+1}^{n+1} x_{ij} \geqslant d_k, \qquad k = 1, 2, \ldots, n. \tag{9.2}$$

To see that the constraints (9.2) describe flows, first rewrite (9.2) as

$$\sum_{i=1}^{k} \sum_{j=k+1}^{n+1} x_{ij} - y_k = d_k, \qquad y_k \geqslant 0. \tag{9.3}$$

Next, successively subtract the $(k-1)$-st equation from the kth, $k = n, n-1, \ldots, 2$, to obtain an equivalent system of constraints. The result is

$$\begin{aligned}
&\sum_{j=2}^{n} x_{ij} - y_1 = d_1,\\
&\ldots\\
&\sum_{j=k+1}^{n} x_{kj} - \sum_{i=1}^{k-1} x_{ik} + y_{k-1} - y_k = d_k - d_{k-1}, \qquad k = 2, \ldots, n,\\
&\ldots\\
&-\sum_{i=1}^{n-1} x_{in} + y_n = -d_n,
\end{aligned} \tag{9.4}$$

subject to which the linear form (9.1) is to be minimized.

The corresponding network is shown in Figure 9.1. Here x_{ij} is

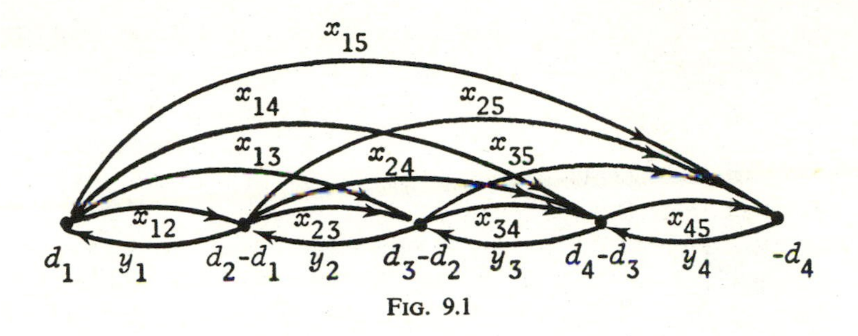

FIG. 9.1

the flow in (i, j) and a_{ij} is the cost per unit of flow; y_i is the flow in $(i + 1, i)$, with $a_{i+1,i} = 0$. Nodes i for which $d_i - d_{i-1} > 0$ are sources with supplies $d_i - d_{i-1}$; nodes i for which $d_i - d_{i-1} < 0$ are sinks with demands $-(d_i - d_{i-1})$.

Referring to Figure 4.2, Part II, and Figure 9.1 above, it is apparent that the equipment replacement problem can be viewed as a special case of capacity scheduling by taking $d_i = 1$, all i.

A number of other situations that can be interpreted in terms of capacity scheduling are described in [41]. Mentioned are models involving checkout and replacement of stochastically failing mechanisms; determination of economic lot sizes, product assortment, and batch-queuing policies; labor-force planning; and multi-commodity warehousing decisions.

One application (the dynamic economic lot size model [43]) deals with the problem described in section 8, where production costs now are concave functions of the number of items produced, and demands must be satisfied. Generally speaking, concavity makes minimization problems difficult, but here it can be seen that it is uneconomical to both produce in a period and carry inventory into the period, and hence there is an optimal policy of the following kind. The total time interval is broken into subintervals, with enough production at the beginning of each of these to satisfy its aggregate demand. Thus finding an optimal policy can be formulated in terms of capacity scheduling by letting a_{ij} be the total cost (including storage) associated with producing enough in period i to satisfy the demands for periods $i, i + 1, \ldots, j - 1$, and by taking all $d_i = 1$. In short, the problem has been reduced to one of finding a cheapest chain from source to sink in the network of Figure 9.1.

10. Minimal spanning trees. A network combinatorial problem for which there is a particularly simple solution method is that of selecting a minimum spanning subtree from an undirected network each of whose links has a length or cost. We may illustrate this problem with the following example. Imagine a number n of cities on a map and suppose that the cost of installing a communication link between cities i and j is $a_{ij} = a_{ji} \geqslant 0$. Each city must be connected, directly or indirectly, to all others, and this is to be done at minimum total cost. Clearly, attention can be confined to trees (acyclic and connected networks of links), for if a connected network contains a cycle, removing one link of the cycle leaves the network connected and reduces cost. A minimal cost tree can be found easily as follows [37]. Select the cheapest link, then the next cheapest, and so on, being sure at each stage that no subset of the selected links forms a cycle. After $n - 1$ selections, a cheapest tree has been constructed.

For example, in the network of Figure 10.1, this procedure might lead to the minimal cost tree shown in heavy links.

While it is not difficult to prove that this method solves the problem, it is nonetheless remarkable that being greedy at each stage works. There are few extremal combinatorial problems for which it does.

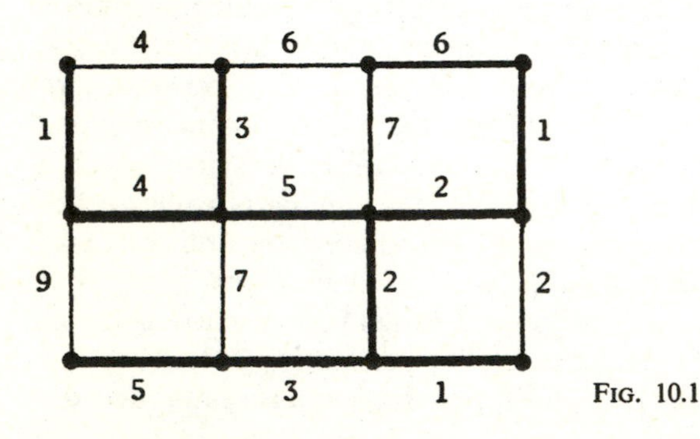

Fig. 10.1

There is an interesting relation between the minimal spanning tree problem and another, which sounds on the surface to be very different. Think of the network in Figure 10.1 as being a highway map, where the number recorded beside each link is the maximum elevation encountered in traversing the link. Suppose someone who plans to drive from i to j dislikes high altitudes and hence wants to find a path connecting i and j that minimizes the maximum altitude. This problem is related to the shortest chain problem in the sense that methods for solving the latter are easily modified to solve the former, and this is so in either the directed or undirected case [40]. But it is also true in the undirected case that the minimal spanning tree solves the problem, and for all pairs of cities. That is, the unique path in the minimal spanning tree joining a pair of cities minimizes the path height [28]. Here we have used "path height" to mean the maximum number on the path.

There is also a min-max theorem concerning paths and cuts for this problem. Call the minimum link number in a cut the "cut height." Then it may be verified that the minimum height of paths joining two nodes is equal to the maximum height of cuts separating the two.

11. The traveling-salesman problem. Many problems that involve minimal connecting networks, and hence superficially resemble that of the last section, have no known simple solution procedures. For example, consider a network connecting a number of cities, with the length of each link being known, and imagine a traveling salesman who must start at some city, visit each of the others just once, and then return to the starting city. How does the salesman determine an itinerary that minimizes the total distance traveled?

A cycle that passes through every node of a network just once is usually called a Hamiltonian cycle. For brevity, we refer to it as a tour. Thus the problem asks for a shortest tour. (Of course the given network may contain no tour, but the existence question can be subsumed by considering all possible links present, those not corresponding to original ones having very large lengths.)

Not a great deal is known, either theoretically or computationally, about this problem, except that it is hard. On the theoretical side, for instance, there seem to be no simple conditions that are necessary and sufficient for a given network to contain a tour. On the computational side, while many methods for determining a shortest tour have been proposed, it is safe to assert that no one of these would guarantee that a problem involving 100 cities, say, could be solved in a reasonable length of time.

12. Minimal k-connected networks. In considering the synthesis of reliable communication networks with respect to link failure, the following question may be raised. Suppose given the complete, undirected network G on n nodes, where again each link of G has an associated number, the cost of installing a communication link between its end-nodes. For each $k = 1, 2, \ldots, n-1$, find a minimal cost k-link-connected spanning subnetwork of G [20]. Here a k-link-connected network is one in which at least k links must be suppressed in order to disconnect the network. Another way of characterizing this property is to say that every pair of nodes is joined by at least k link-disjoint paths. Thus k might be thought of as the "reliability level" of the communication network, and the practical problem is to minimize cost while achieving a stipulated reliability level.

For $k = 1$, the problem is that of section 10, and hence is readily solved. For $k = 2$ and all link costs 1 or ∞, the problem becomes that of determining whether a given network (the subnetwork of unit cost links) contains a tour, and is thus already difficult. But if all link costs are equal, the answer is known. Here the problem is to determine the minimum number of links required in a k-link-connected network on n nodes. For $k \geqslant 2$, there is an obvious lower bound on the number needed, namely $kn/2$ (for even kn) or $(kn + 1)/2$ (for odd kn). These bounds can always be achieved.

REFERENCES

1. Busacker, R. G., and P. J. Gowen, "A procedure for determining a family of minimal cost network flow patterns," *O.R.O. Technical Paper*, **15** (1961).
2. Cartaino, T. F., and S. E. Dreyfus, "Application of dynamic programming to the airplane minimum time-to-climb problem," *Aero. Engr. Rev.*, **16** (1957), 74–77.
3. Dantzig, G. B., "Application of the simplex method to a transportation problem," *Activity Analysis of Production and Allocation*, Cowles Commission Monograph **13**, New York: Wiley, 1951, 359–373.
4. Dantzig, G. B., and D. R. Fulkerson, "Minimizing the number of tankers to meet a fixed schedule," *Naval Res. Logist. Quart.*, **1** (1954), 217–222.
5. ——, "On the max-flow min-cut theorem of networks," *Linear Inequalities and Related Systems*, Ann. of Math. Study **38**, Princeton Univ. Press, (1956), 215–221.
6. Dantzig, G. B., D. R. Fulkerson, and S. Johnson, "Solution of a large scale traveling salesman problem," *Op. Res.*, **2**, 393–410.
7. Derman, C., and M. Klein, "A note on the optimal depletion of inventory," *Management Sci.*, **5** (1959), 210–214.
8. Dilworth, R. P., "A decomposition theorem for partially ordered sets," *Ann. of Math.*, **51** (1950), 161–166.
9. Dreyfus, S. E., "A generalized equipment replacement study," *J. Soc. Indust. Appl. Math.*, **8** (1960), 425–435.
10. Ford, L. R., Jr., "Network flow theory," *The Rand Corp.*, *P*-**923** (1956).
11. Ford, L. R., Jr., and D. R. Fulkerson, "Maximal flow through a network," *Canad. J. Math.*, **8** (1956), 399–404.
12. ——, "A simple algorithm for finding maximal network flows and an application to the Hitchcock problem," *Canad. J. Math.*, **9** (1957), 210–218.
13. ——, "A suggested computation for maximal multi-commodity network flows," *Management Sci.*, **5** (1958), 97–101.

14. ——, "Constructing maximal dynamic flows from static flows," *Op. Res.*, **6** (1958), 419–433.

15. ——, *Flows in Networks*, Princeton Univ. Press, 1962, 194 p.

16. Fulkerson, D. R., "Increasing the capacity of a network: The parametric budget problem," *Management Sci.*, **5** (1959), 472–483.

17. ——, "An out-of-kilter method for minimal cost flow problems," *J. Soc. Indust. Appl. Math.*, **9** (1961), 18–27.

18. ——, "A network flow computation for project cost curves," *Management Sci.*, **7** (1961), 167–178.

19. ——, "A network flow feasibility theorem and combinatorial applications," *Canad. J. Math.*, **11** (1959), 440–451.

20. Fulkerson, D. R., and L. S. Shapley, "Minimal k-arc-connected graphs," *The Rand Corp., P-***2371** (1961), 11 p.

21. Gale, D., "A theorem on flows in networks," *Pac. J. Math.*, **7** (1957), 1073–1082.

22. Gomory, R. E., and T. C. Hu, "Multi-terminal network flows," *J. Soc. Indust. Appl. Math.*, **9** (1961), 551–571.

23. Gorham, W., "An application of a network flow model to personnel planning," *The Rand Corp., RM-***2587** (1960), 85 p.

24. Hall, P., "On representatives of subsets," *J. London Math. Soc.*, **10** (1935), 26–30.

25. Hitchcock, F. L., "The distribution of a product from several sources to numerous localities," *J. Math. and Phys.*, **20** (1941), 224–230.

26. Hoffman, A. J., "Some recent applications of the theory of linear inequalities to extremal combinatorial analysis," *Proc. Symposia Applied Math.*, **10** (1960).

27. Hu, T. C., "Parallel sequencing and assembly line problems," *Op. Res.*, **9** (1961), 841–849.

28. ——, "The maximum capacity route problem," *Op. Res.*, **9** (1961), 898–900.

29. Jewell, W. S., "Optimal flow through networks with gains," *Proc. Second International Conf. on Oper. Res.*, Aix-en-Provence, France, 1960.

30. Johnson, S. M., "Sequential production planning over time at minimum cost," *Man. Sci.*, **3** (1957), 435–437.

31. ——, "Optimal two- and three-stage production schedules with setup

times included," *Naval Res. Log. Q.*, **1** (1954), 61–68.

32. Kantorovitch, L., and M. K. Gavurin, "The application of mathematical methods in problems of freight flow analysis," *Collection of Problems Concerned with Increasing the Effectiveness of Transports*, Publ. of the Akad. Nauk SSSR, Moscow-Leningrad: 1949, 110–138.

33. Kelley, J. E., "Critical path planning and scheduling: mathematical basis," *Op. Res.*, **9** (1961), 296–321.

34. Koopmans, T. C., and S. Reiter, "A model of transportation," *Activity Analysis of Production and Allocation*, Cowles Commission Monograph **13**, New York: Wiley, 1951, 229–259.

35. König, D., *Theorie der endlichen und unendlichen Graphen*, Chelsea: 1950, 258 p.

36. Kuhn, H. W., "The Hungarian method for the assignment problem," *Naval Res. Log. Q.*, **2** (1955), 83–97.

37. Kruskal, J. B., Jr., "On the shortest spanning subtree of a graph and the traveling salesman problem," *Proc. Amer. Math. Soc.*, **7** (1956), 48–50.

38. Minty, G. J., "Monotone networks," *Proc. Roy. Soc.*, A, **257** (1960), 194–212.

39. Orden, A., "The transshipment problem," *Man. Sci.*, **3** (1956), 276–285.

40. Pollack, M., "The maximum capacity route through a network," *Op. Res.*, **8** (1960), 733–736.

41. Veinott, A. F., Jr., and H. M. Wagner, "Optimal capacity scheduling —I and II," *Op. Res.*, **10** (1962), 518–547.

42. Votaw, D. F., Jr., and A. Orden, "The personnel assignment problem," *Project SCOOP, Manual* **10** (1952), 155–163.

43. Wagner, H. M., and T. M. Whitin, "Dynamic version of the economic lot-size model," *Man. Sci.*, **5** (1958), 89–96.

MULTI-TERMINAL FLOWS IN A NETWORK

*R. E. Gomory and T. C. Hu**

First, let us distinguish the words "graph" and "network." A graph G is defined as a set of points (vertices, nodes) connected by lines (edges, branches, arcs). A network can be thought as a graph with numbers associated with the nodes and arcs. The meanings of those numbers depend on the specific case of application. In an electrical network, the numbers associated with the lines may be the resistances of the wires and the numbers associated with the points may be the potentials at these points.

Since 1956, there has been a growing number of papers dealing with network flow theory. This is not to be confused with the electrical network theory which was first started by Kirchhoff in 1847. Network flow problems were first considered by Ford and Fulkerson [3] and the interested reader is referred to the books by Ford and Fulkerson [5] and Hu [8]. In the present paper, we

*The work of the second author is sponsored by the United States Army under Contract No.: DA-31-124-ARO-D-462 and National Science Foundation Grant G. J. 28339.

consider a special network flow problem which has also been considered by Mayeda [10]. This paper is essentially the same as the paper, "Multi-terminal Network Flows," by Gomory and Hu [6] but with some introductory material added to make this paper self-contained.

We consider a network N consisting of nodes N_i ($i = 1, 2, \ldots, n$) and arcs B_{ij} connecting the nodes N_i and N_j. Each arc B_{ij} has associated with it a nonnegative number b_{ij} called the *arc capacity*. Throughout the paper, we assume that there are a finite number of nodes, that the network is connected, and that the arcs have no orientation, hence $b_{ij} = b_{ji}$. The arc capacity b_{ij} may be thought of intuitively as representing the maximal amount of some commodity that can go through the arc from N_i to N_j. For example, the commodity may be "water" and the arc capacity may indicate the cross-sectional area of the pipeline. Or the commodity may be cars in highways, and b_{ij} indicates that the highway between city i and city j is two-lane or four-lane. We shall consider the case that there are two distinguished nodes N_s and N_t, called the source and the sink, respectively.

The source is the entry of flow to the network, and the sink is the exit of flow from the network. Mathematically, we define the *arc flows* x_{ij}, and in particular the flows x_{sj} from the source and x_{it} into the sink, as a set of nonnegative numbers satisfying the following constraints:

$$\sum_i x_{ij} - \sum_k x_{jk} = \begin{cases} -v & \text{if} \quad j = s \\ 0 & \text{if} \quad j \neq s, t \\ v & \text{if} \quad j = t \end{cases} \tag{1}$$

$$0 \leqslant x_{ij} \leqslant b_{ij} \quad \text{for all} \quad i, j. \tag{2}$$

Note that flow is conserved at every node except the source and the sink, and each arc flow x_{ij} is bounded from above by b_{ij}, the arc capacity. The *value* of the flow from N_s to N_t is v. A fundamental question in network flow theory is to find the maximal value of v, given the b_{ij} of the network. To answer the question, let us describe the notion of a cut in a network. A cut is denoted by $(A, \bar{A})$ where A is a subset of nodes of the network and $\bar{A}$ is its complement. A cut $(A, \bar{A})$ is the set of arcs connecting a node in A with a node in $\bar{A}$. Thus, a cut is a set of arcs, the removal of which will disconnect the network. A cut separating N_s and N_t is a cut $(A, \bar{A})$ with $N_s \in A$ and $N_t \in \bar{A}$. The *capacity of a cut*, denoted by $b_{A,\bar{A}}$, is Σb_{ij} where $N_i \in A$ and $N_j \in \bar{A}$. Due to the constraints (1) and (2), the maximal flow value v from N_s to N_t certainly cannot exceed the capacity of any cut separating N_s and N_t. The cut with the minimal capacity is called a minimal cut. Now we can state without proof the well-known max-flow min-cut theorem. (See Ford and Fulkerson [3] [4] [5].)

THEOREM 1: *The maximal flow value from the source to the sink is equal to the capacity of a minimal cut separating the source and the sink.*

There is a constructive proof of Theorem 1 and an algorithm to locate a minimal cut, but we shall not discuss them here. If the b_{ij} are positive integers and we interpret them as the number of edges connecting node i and node j, then the minimal cut is the minimum number of edges to be deleted to disconnect the graph into two components, one containing N_s and the other containing N_t. And the maximal flow value v is the maximal number of edge-disjoint paths between N_s and N_t.

Given an n-node network with $b_{ij} = b_{ji}$, there are $\binom{n}{2}$ possible choices of pairs of nodes to be the source and the sink. For each pair of nodes i and j selected, we can use the max-flow min-cut theorem to find the maximal flow value f_{ij}. Thus we can get a symmetric matrix of size $n \times n$, where the (i, j) entry denotes the maximal flow value f_{ij} between N_i and N_j, for $i \neq j$. It is convenient to take $f_{ii} = \infty$.

Consequently, for each network there are two associated symmetric matrices: The matrix B of the b_{ij} and the matrix F of the resulting flows f_{ij}. Not any matrix can be an f_{ij} matrix, so it is natural to ask when a given set of flow values f_{ij} can be realized by some network. One answer in terms of the ability to repeatedly partition the matrix $F = (f_{ij})$ in a particular way has already been given by Mayeda [10]. Here we give another necessary and sufficient condition—a sort of "triangle inequality."

This condition reduces the problem of deciding whether or not a given matrix F is realizable by some network to the well-known problem of constructing a maximal tree of a network—a problem already solved in a very effective manner first by Kruskal [9] and even more efficiently by Prim [11].

THEOREM 2: *A necessary and sufficient condition for a matrix F to be realizable is that*

$$f_{ik} \geqslant \min(f_{ij}, f_{jk}) \tag{3}$$

for all i, j, k.

Proof: We first show necessity. Suppose the theorem were false and for some i, j, k, $f_{ik} < \min(f_{ij}, f_{jk})$. Then there exists by the maximum-flow minimum-cut theorem [3] a cut or division of the nodes into two sets A and $\bar{A}$ with $N_i \in A$ and $N_k \in \bar{A}$ such that the sum of the capacities of the arcs connecting nodes in A with nodes in $\bar{A}$ is f_{ik}. Now N_j belongs either to A or $\bar{A}$. If it is in A, then it is cut off from N_k by the cut; since the capacity of the cut is $< f_{jk}$, this is a contradiction. Similarly N_j cannot be in $\bar{A}$, for then it is cut off from N_i. Therefore, $f_{ik} \geqslant \min(f_{ij}, f_{jk})$.

Once established, the relation $f_{ik} \geqslant \min(f_{ij}, f_{jk})$ has, by induction, the immediate consequence

$$f_{ip} \geqslant \min(f_{ij}, f_{jk}, f_{kl}, \ldots, f_{op}) \tag{4}$$

where $N_i, N_j, \ldots, N_p$ is any sequence of nodes.

Now for the sufficiency. For this we need first the notion of spanning tree, then the notion of maximal spanning tree. The notion of tree we take as known. (See, for example, [1].) A spanning tree is simply a tree that includes all nodes. If there are numbers n_{ij} attached to the arcs of a tree, one may introduce the value of a tree as the sum of the numbers n_{ij} on the arcs of the tree.

Among spanning trees, there is one or more whose value is

maximal among spanning trees. This is a maximal spanning tree and can easily be constructed by Prim's method. Any maximal spanning tree has the following easily established property. Let N_i and N_p be two nodes whose connecting arc is not in the tree; then the number of that connecting arc satisfies $n_{ip} \leqslant \min(n_{ij}, n_{jk}, \ldots, n_{op})$ where the $n_{ij}, \ldots, n_{op}$ are numbers on the arcs of the (unique) path connecting N_i to N_p in the tree. For if the inequality did not hold, the smallest arc in the tree path could be removed and the arc joining N_i and N_p substituted to form a tree with value larger than the maximum.

We can construct a maximal spanning tree on the complete graph of n nodes using f_{ij} as attached numbers to the arcs connecting N_i and N_j. From the maximality we have as above for a direct arc being compared with a path through the tree, $f_{ip} \leqslant \min(f_{ij}, \ldots, f_{op})$, while from (4) we have the opposite inequality. So for any arc not in the tree

$$f_{ip} = \min(f_{ij}, \ldots, f_{op}). \tag{5}$$

However, this is precisely the flow which results if a network is constructed with branch capacity $b_{ij} = f_{ij}$ for arcs in the tree, and $b_{ij} = 0$ otherwise. Thus any F matrix satisfying the condition is realizable. This ends the proof.

To see if a given matrix is in fact realizable, one could construct a maximal spanning tree using Prim's method, then check to see if condition (3) is satisfied. Actually, it is far more economical (and extremely easy) to check (3) during the course of Prim's algorithm as described in the appendix.

We next turn to the problem of analysis, i.e., given a network of arcs B_{ij} with capacities b_{ij}, what are the resulting maximal flow values f_{ij}? We have seen from (5) that any flow value is numerically equal to some flow value in the maximal spanning tree. As there are only $n - 1$ arcs in a spanning tree (where n is the number of nodes), there are only $n - 1$ numerically different flow values possible. This makes it reasonable to suspect that all $n(n - 1)/2$ flows f_{ij} can be obtained by something better than doing $n(n - 1)/2$ flow problems. We will, in fact, show that all flows can be deduced after only $n - 1$ flow problems have been computed.

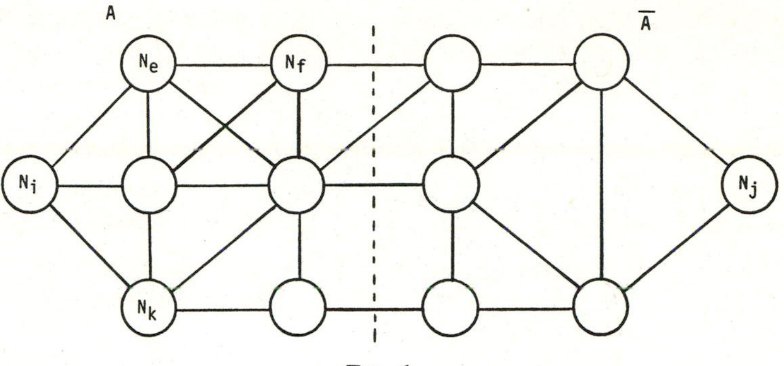

FIG. 1

Consider first a network such as the one shown in Figure 1 with a minimal cut $(A, \bar{A})$ separating N_i and N_j, $N_i \in A$, $N_j \in \bar{A}$.

Then let us construct a slightly different network, one in which all nodes in $\bar{A}$ are replaced by a single special node P to which all the arcs of the cut are attached (we can replace several arcs connecting the same two nodes by one arc having the total capacity). In this condensed network (Figure 2), consider the maximal flow between two ordinary nodes, N_e and N_k. We shall show

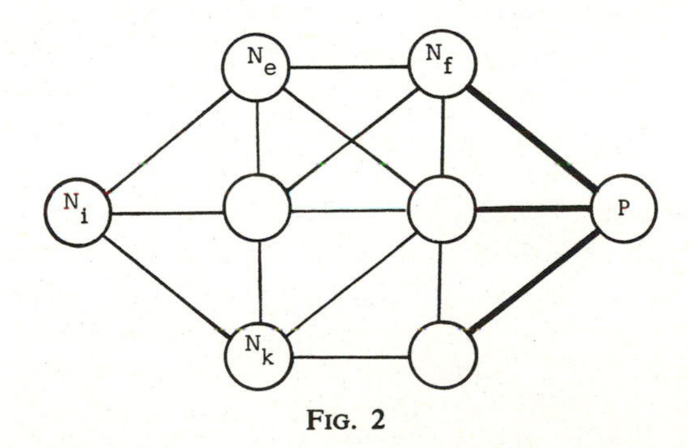

FIG. 2

LEMMA 1: *The flow value between two ordinary nodes N_e and N_k in the condensed network is numerically equal to the flow f_{ek} in the original network.*

Proof: Let $(B, \bar{B})$ be a minimal cut separating N_e and N_k in the original network and define sets of nodes

$$X = A \cap B, \qquad \bar{X} = A \cap \bar{B},$$

$$Y = \bar{A} \cap B, \qquad \bar{Y} = \bar{A} \cap \bar{B}.$$

Here $\bar{X}$ is the complement of X in A, $\bar{Y}$ is the complement of Y in $\bar{A}$. We may assume that $N_e \in X$, $N_k \in \bar{X}$ and $N_i \in X$. Let $b_{XY} = \Sigma b_{ij}$ where $N_i \in X$ and $N_j \in Y$.

CASE 1. $N_j \in Y$. Now

$$b_{A\bar{A}} = b_{XY} + b_{X\bar{Y}} + b_{\bar{X}Y} + b_{\bar{X}\bar{Y}},$$

$$b_{B\bar{B}} = b_{X\bar{X}} + b_{X\bar{Y}} + b_{\bar{X}Y} + b_{Y\bar{Y}}.$$

Since $(B, \bar{B})$ is a minimal cut separating N_e and N_k, and since $(X \cup Y \cup \bar{Y}, \bar{X})$ separates N_e and N_k, we have

$$b_{B\bar{B}} - b_{X \cup Y \cup \bar{Y}, \bar{X}} = b_{X\bar{Y}} + b_{Y\bar{Y}} - b_{\bar{X}\bar{Y}} \leqslant 0. \tag{6}$$

Since $(A, \bar{A})$ is a minimal cut separating N_i and N_j, and since $(X \cup \bar{X} \cup \bar{Y}, Y)$ separates N_i and N_j, then

$$b_{A\bar{A}} - b_{X \cup \bar{X} \cup \bar{Y}, Y} = b_{X\bar{Y}} + b_{\bar{X}\bar{Y}} - b_{Y\bar{Y}} \leqslant 0. \tag{7}$$

Adding (6) and (7) shows that $b_{X\bar{Y}} \leqslant 0$ and hence $b_{X\bar{Y}} = 0$. It then follows from (6) and (7) that $b_{Y\bar{Y}} - b_{\bar{X}\bar{Y}} = 0$ also. Hence $(X \cup Y \cup \bar{Y}, \bar{X}) = (X \cup \bar{A}, \bar{X})$ is also a minimal cut separating N_e and N_k.

CASE 2. $N_j \in \bar{Y}$. A similar proof shows that $(X, \bar{X} \cup \bar{A})$ is a minimal cut separating N_e and N_k in this case.

In other words, there is always a minimal cut separating N_e and N_k such that the set of nodes $\bar{A}$ is on one side of this cut. Since the flow value is determined by the value of this minimum cut, which is unchanged by the condensing process, condensing $\bar{A}$ to a single node does not increase the value of a maximal flow from N_e to N_k. On the other hand, condensing process certainly cannot decrease the maximal flow value between N_e and N_k.

Thus any flow between N_e and N_k in the condensed network gives rise to an equal flow in the original network. This completes the proof.

Since a cut in the condensed network gives a cut in the original, and the maximal flow values are the same, a minimal cut in the condensed network gives, simply by replacing P by $\bar{A}$, a minimal cut in the original.

We now proceed to the analysis.

One procedure is simply this. We take two nodes and do a maximal flow computation [4] to find a minimal cut $(A, \bar{A})$. We represent this by two generalized nodes connected by an arc bearing the cut value (Figure 3). In one node are listed the nodes of A, in the other those of $\bar{A}$. We now repeat this process. Choose two nodes in A (or two in $\bar{A}$), and solve the flow problem in the condensed network in which $\bar{A}$ (or A) is a single node. The resulting cut has a value v_2 and is represented by a link connecting the two parts into which A is divided by the cut, say A_1 and A_2. $\bar{A}$ is attached to A_1 if it is in the same part of the cut as A_1, to A_2 if it is in the same part as A_2 (Figure 4).

The cutting is then continued. At each stage we have certain generalized nodes (which may represent many nodes of the ori-

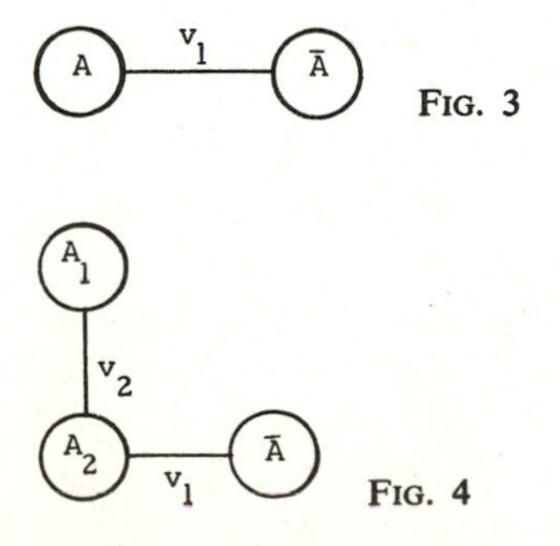

FIG. 3

FIG. 4

ginal network), and certain arcs connecting them. To proceed with the computation, we select a generalized node A_i and two original nodes N_a and N_b in A_i. Upon the removal of all arcs which connect to A_i, the network of generalized nodes falls into a number of disconnected components. We condense each component, except A_i itself, into a single node and solve the network flow problem consisting of these condensed nodes and the original nodes within A_i, and using N_a and N_b as source and sink. The minimal cut obtained by this flow calculation splits A_i into two parts, A_{i1}, A_{i2}. This is represented in the diagram by replacing A_i by two generalized nodes A_{i1} and A_{i2} connected by an arc bearing the cut value. All other arcs and generalized nodes in the diagram are unchanged except those arcs which formerly connected to A_i. Such an arc is now attached to A_{i1} if its component was on the same side of the cut as the nodes in A_{i1}, and attached to A_{i2} if its component fell on the other side.

This process is repeated until the generalized nodes of the diagram consist of exactly one node each. This point is reached after exactly $n - 1$ cuts, for the diagram is a tree at all times, so when the process stops it is an n-node tree and so has $n - 1$ branches, each created by solving a flow problem in a network equal to or smaller in size than the original.

We then assert

LEMMA 2: *The flow value between any two nodes is simply*

$$\min(v_{ij}, v_{jk}, \ldots, v_{qr}),$$

where the v_{ij} are values of a series of arcs of the tree connecting the two nodes.

Before proceeding to prove this last assertion, it is probably a good idea to illustrate the process by an example:

Taking as our B_{ij} network the net in Figure 5, we arbitrarily choose nodes 2 and 6, and upon doing a flow problem we find the minimum cut to be (as indicated in Figure 5) (1, 2, |3, 4, 5, 6) with

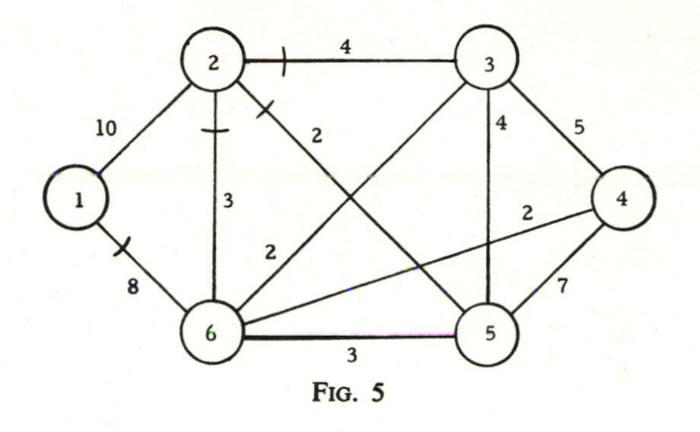

FIG. 5

capacity 17. This is represented by the diagram

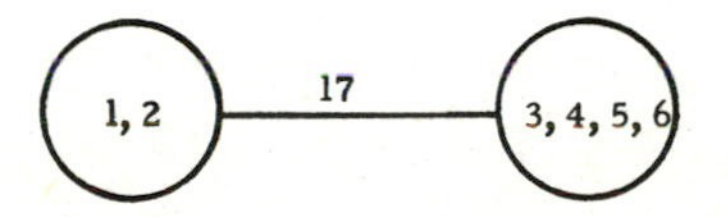

To get the flow 1-2, we consider 3, 4, 5, 6 as a single node, obtaining Figure 6, in which the minimum cut 1-2 is (1|2, 3, 4, 5, 6) with capacity 18, so we obtain the diagram

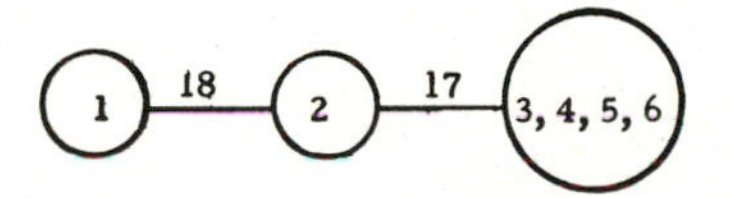

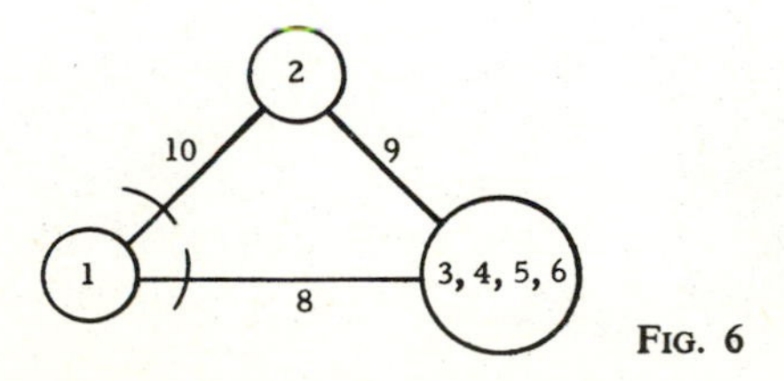

FIG. 6

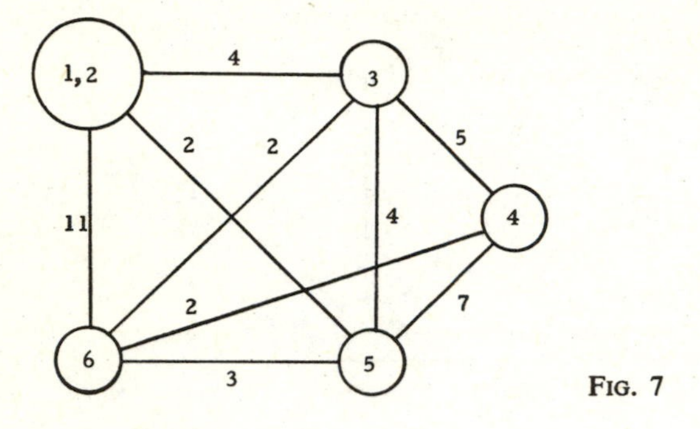

FIG. 7

We next choose 3 and 6. Considering 1 and 2 as a single node (Figure 7), we find the minimum cut (1, 2, 6|3, 4, 5), capacity 13. So the diagram becomes

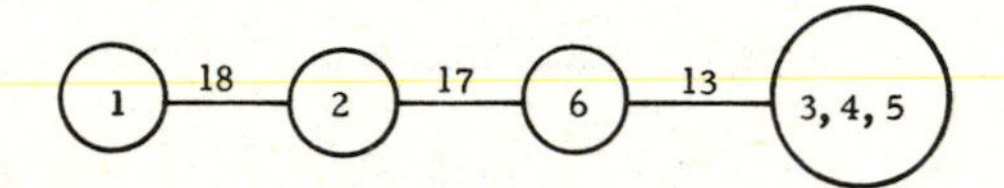

We next consider the flow 4-5, taking 1, 2, 6 as a node (Figure 8), the resulting minimum cut being (4|1, 2, 3, 5, 6) with capacity 14. So we have

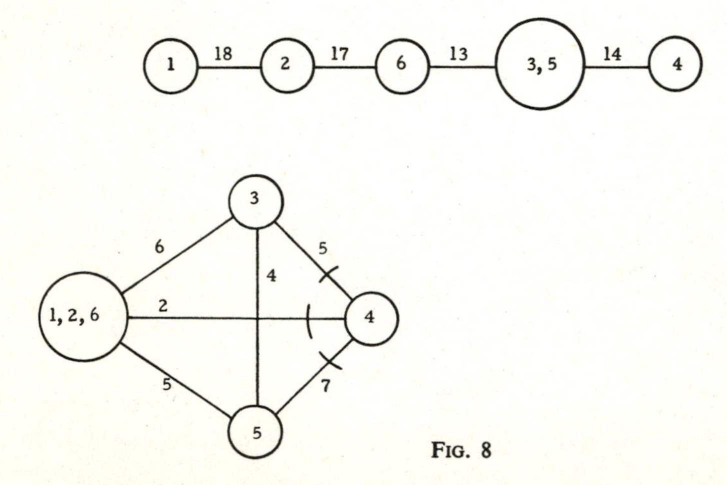

FIG. 8

Finally, we consider the flow 3-5, taking 1, 2, 6 as one node and 4 as the other to get the same network as above,

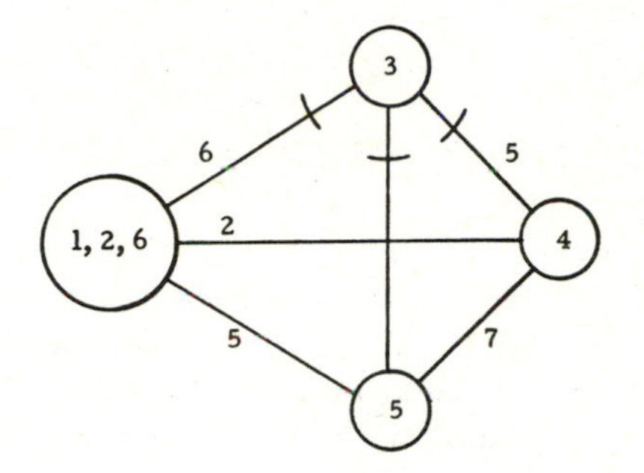

the minimum cut 3-5 being (3|1, 2, 4, 5, 6) with capacity 15, giving as the final tree,

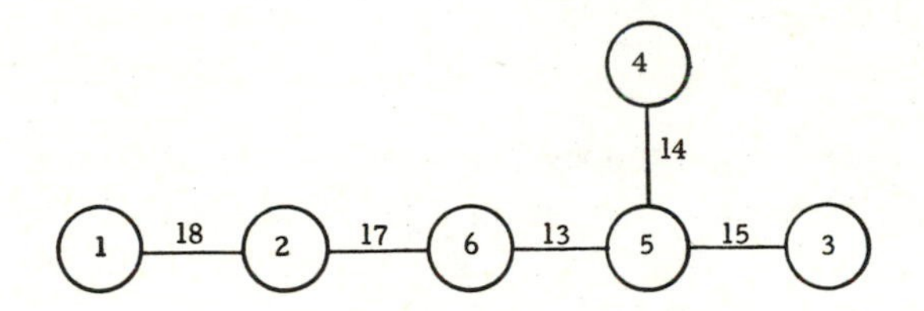

We would now assert that the maximum flow from 1 to 3, f_{13}, is 13, the maximum flow f_{16} is 17, etc.

We now proceed to prove Lemma 2. Consider two nodes N_i and N_j. We certainly have

$$f_{ij} \leqslant \min(v_1, \ldots, v_r),$$

for each v_i on the path connecting N_i and N_j corresponds to a cut separating N_i and N_j. To show the reverse inequality is a little more difficult. Consider any stage of the construction

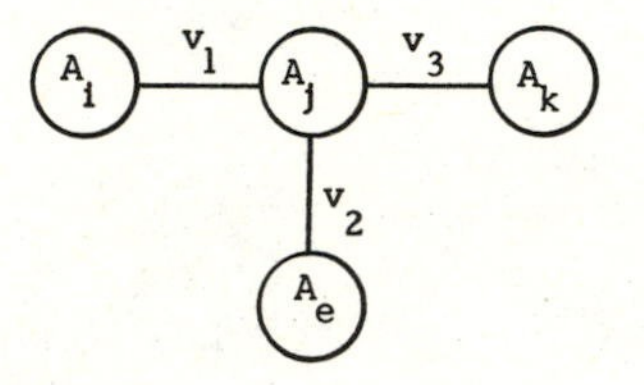

where we have arcs representing cuts and nodes representing sets. We assert that if an arc of value v connects sets A_i and A_j then there is a node N_i in A_i and a node N_j in A_j such that $f_{ij} = v$.

This is certainly true after the first cut. We will show that the property is maintained. Consider an A_i about to be cut,

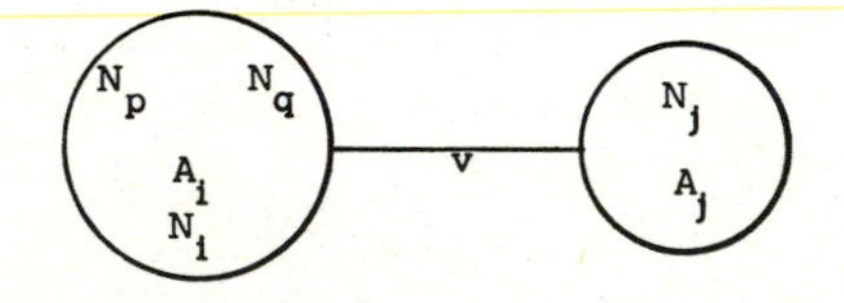

with A_j representing the set connected to A_i by the arc v. By the induction hypothesis there is an N_i in A_i and N_j in A_j with $f_{ij} = v$. After cutting N_p from N_q, A_i divides into A_{i_p} and A_{i_q}. We can assume A_j is attached to A_{i_p}.

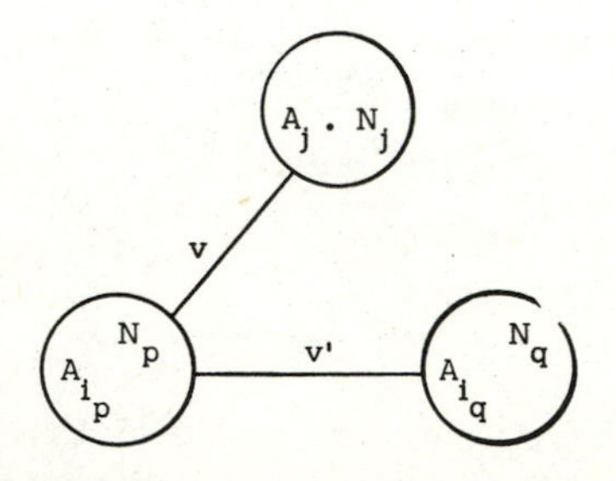

Clearly N_p and N_q provide the desired flow $f_{pq} = v'$ across the new link. As to the old link of value v there are two cases:

(i) $N_i \in A_{i_p}$.

Then the flow $f_{ij} = v$ is still applicable.

(ii) $N_i \in A_{i_q}$.

Then consider the nodes N_j, N_i, N_q, and N_p. From (4)

$$f_{jp} \geqslant \min(f_{ji}, f_{iq}, f_{qp}).$$

Since N_j and N_p are on one side of the cut whose value is v' and N_i and N_q are on the other, we know that the flow f_{jp} is unaffected if A_{i_q} is replaced by a single node or, what is the same thing, if all arcs within A_{i_q} are given an arbitrarily large capacity M. Doing this makes f_{iq} large so we have

$$f_{jp} \geqslant \min(f_{ji}, f_{qp}).$$

Since $f_{ji} = v$ and $f_{qp} = v'$, we have

$$f_{jp} \geqslant \min(v, v').$$

Since the cut separating N_j and N_i is of value v', we must have $v' \geqslant f_{ij} = v$. So finally $f_{jp} \geqslant v$. As v is the value of a cut separating N_j and N_p, this implies $f_{jp} = v$. Thus N_j and N_p provide the two needed nodes.

Since we now know that in the final tree the values on the links actually represent flow values between the adjacent points, the reverse inequality

$$f_{ij} \geqslant \min(v_1, \ldots, v_r)$$

is once again just an application of (4).

This establishes the desired result

$$f_{ij} = \min(v_1, \ldots, v_r).$$

Hence the flow matrix for our example is

	1	2	3	4	5	6
1	∞	18	13	13	13	17
2	18	∞	13	13	13	17
3	13	13	∞	14	15	13
4	13	13	14	∞	14	13
5	13	13	15	14	∞	13
6	17	17	13	13	13	∞

In any tree diagram an arc from N_i to N_j can be considered as representing a cut of the nodes, since its removal divides the nodes of the tree into two sets, A and $\bar{A}$. If, in addition, each cut so obtained is a minimal cut between N_i and N_j in some network N, and the attached branch value in the tree is the capacity of the cut in N, then the tree is called a *cut tree* of N.

We have just shown a way of obtaining a cut tree of a network by solving $n - 1$ flow problems.

If we are interested in maximal flows between p nodes where $2 \leqslant p \leqslant n$, then only $p - 1$ flow problems are needed. This generalization is discussed in [8]. The procedure is essentially the same. We stop the computation when each generalized node contains only one of the p nodes which are of interest to us.

Given an $n \times n$ matrix of nonnegative numbers $r_{ij} (r_{ij} = r_{ji})$, we shall call an n-node network N *satisfactory if its flows satisfy*

$$f_{ij} \geqslant r_{ij} \qquad (\text{all } i, j;\ i \neq j).$$

The synthesis problem we consider is the one of finding a satisfactory network having smallest cost. If we say that the cost of installing one unit of branch capacity between N_i and N_j is c_{ij}, then it seems necessary to use the apparatus of linear programming and this approach is developed in [7]. However, if we try to find the satisfactory network of least total branch capacity, this is equivalent to the case when $c_{ij} = 1$ for all $i, j, i \neq j$. Then special methods can be devised, and it is this problem we take up now.

We first introduce a tree T of dominant requirements. This is simply any maximal spanning tree constructed using the r_{ij} as arc values. For example, given the requirements

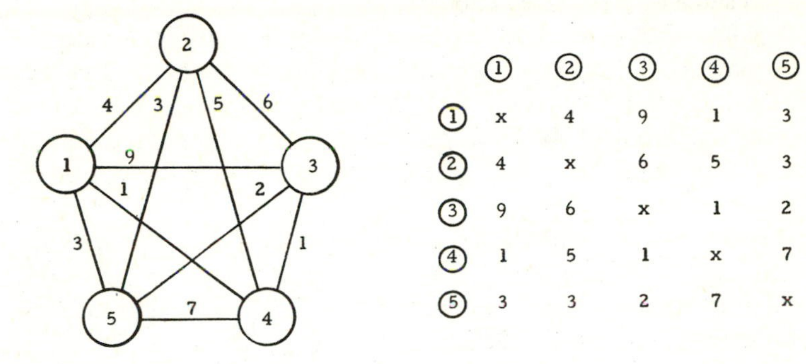

	①	②	③	④	⑤
①	x	4	9	1	3
②	4	x	6	5	3
③	9	6	x	1	2
④	1	5	1	x	7
⑤	3	3	2	7	x

an easily constructed dominant requirement tree is

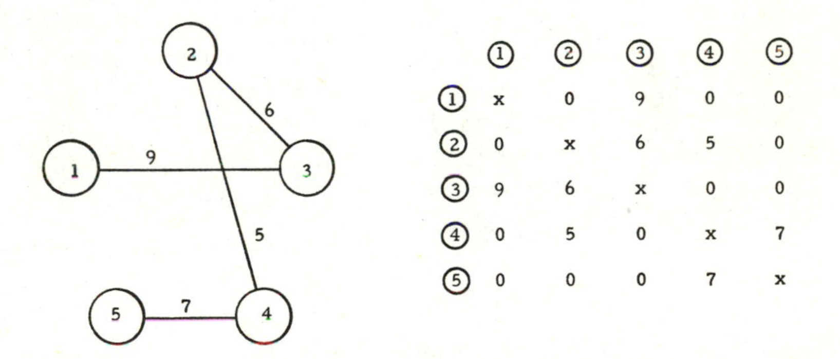

	①	②	③	④	⑤
①	x	0	9	0	0
②	0	x	6	5	0
③	9	6	x	0	0
④	0	5	0	x	7
⑤	0	0	0	7	x

Any satisfactory network must of course satisfy all requirements $f_{ij} \geqslant r_{ij}$ where r_{ij} are attached to arcs in T. This is also sufficient since the missing r_{ip} satisfy the usual relation

$$r_{ip} \leqslant \min(r_{ij}, r_{jk}, \ldots, r_{qp}),$$

where the r's on the right hand side form the unique path in T. In any network satisfying the *dominant tree requirements* (*dominant requirements* for short), the flow f_{ip} must automatically satisfy

$$f_{ip} \geqslant \min(f_{ij}, f_{jk}, \ldots, f_{qp}) \geqslant \min(r_{ij}, \ldots, r_{qp}) \geqslant r_{ip},$$

and so satisfy all requirements.

Because of this, we shall henceforth consider only the dominant requirements. This is not necessary, only convenient: the methods we shall develop can easily be modified to apply directly to the original requirement network, but the exposition is simplified and essentials brought out more easily by working with the dominant requirements.

We define the total branch capacity to be $\frac{1}{2}\Sigma_{i\neq j} b_{ij}$. We now introduce a lower bound C_L for this quantity. Consider any node N_i of the network. Let $u_i = \max_j r_{ij}$, that is, u_i is the largest flow requirement out of N_i. Define $C_L = \frac{1}{2}\Sigma_i u_i$. Then as any satisfactory network N must provide capacities b_{ij} capable of carrying this flow out of N_i, $\Sigma_j b_{ij} \geqslant u_i$, and thus the total branch capacity $= \frac{1}{2}\Sigma_{i\neq j} b_{ij} \geqslant \frac{1}{2}\Sigma_i u_i = C_L$. This bound is due to Chien [2].

The number C_L can also be computed directly from T without reverting to the original requirement, since the same u_i results if the $\max_j r_{ij}$ is taken only over branches of T adjacent to N_i.

Now consider a fixed tree T with attached numbers r_{ij} and resulting bound C_L. If the r_{ij} are replaced by a different set r'_{ij} a new bound C'_L results. If we use $r''_{ij} = r'_{ij} + r_{ij}$ on the arcs of the same fixed tree we get C''_L always with

$$C''_L \leqslant C'_L + C_L,$$

but if r_{ij} (or r'_{ij}) are "uniform" requirements, i.e., $r_{ij} = \beta$ for all r_{ij} in T, then, obviously, equality holds, so

$$C''_L = C'_L + C_L, \tag{9}$$

a result we will use in what follows.

One further remark is needed. If a branch capacity network B_{ij} with capacities b_{ij} and flows f_{ij} is superposed on another having the same nodes, arcs, and different capacities b'_{ij} and flows f'_{ij}, then the resulting network is taken as having arcs with capacities

$b''_{ij} = b'_{ij} + b_{ij}$. The new flows f''_{ij} clearly satisfy

$$f''_{ij} \geqslant f'_{ij} + f_{ij}. \tag{10}$$

Consider a requirement tree T with $n - 1$ requirements which we will now designate. Let the smallest be $r_{\min}$. Since all requirements can be written as $r_{\min} + (r_{ij} - r_{\min})$, we can regard T as being obtained by superposing a uniform requirement tree T with uniform requirement $r_{\min}$ on two smaller trees with requirements $r_{ij} - r_{\min}$. For example, the dominant requirement tree in the preceding example can be regarded as the superposition of a uniform 5-tree, i.e., dominant requirement tree with $r_{ij} = 5$ (Figure 9(a) and the two smaller trees in Figure 9(b) and 9(c)).

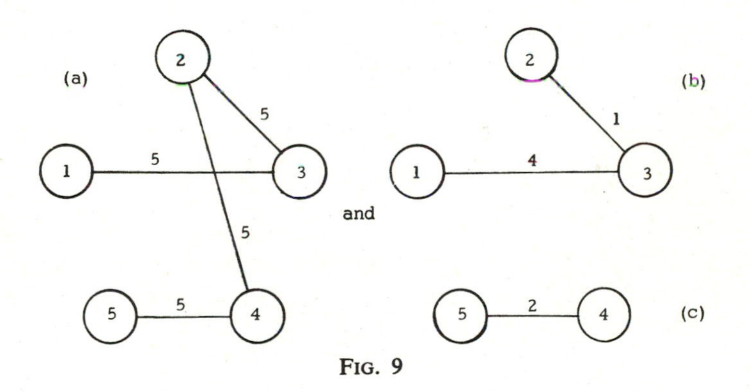

FIG. 9

If the uniform tree and the residual parts can be synthesized by networks N_a, N_b, N_c, so that their individual C_L's are actually attained, then by simply superposing the synthesized networks we get new flows which by (10) equal or exceed the requirements, while the total capacity used is by (9) equal to the lowest possible amount C_L. From now on, we shall use the phrase "synthesizing trees" to mean "constructing a network with maximum flows greater or equal to the requirements in the tree."

The synthesis problem for T then has been reduced to the synthesis of smaller trees and a uniform requirement tree, such that their lower bounds C_L are actually obtained.

However, we can repeat the decomposition process on the smaller trees until only uniform trees remain so that the problem is actually reduced to synthesizing uniform trees.

This however is extremely easy. Given any tree T' with uniform requirement β, the lower bound C_L is $n\beta/2$ and a suitable network is constructed by drawing any cycle through the nodes and then assigning capacity $\beta/2$ to each arc of the cycle. (In the smallest case, $n = 2$, both links of the cycle coincide and a single arc of capacity β is used.)

In the case of our example, to carry out the process we continue the decomposition begun in Figure 9 by decomposing (b) further so that the tree T becomes the sum of

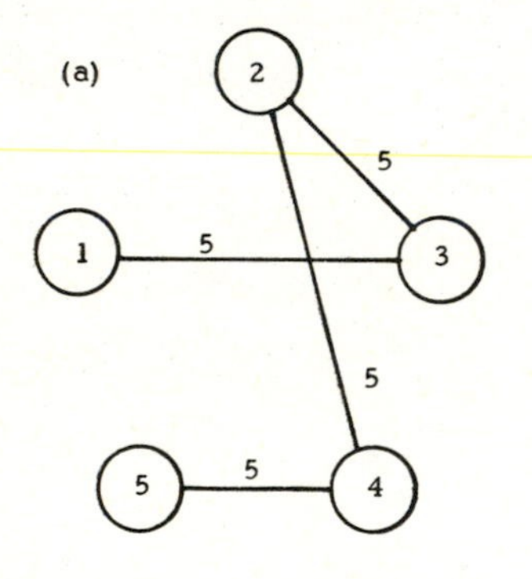

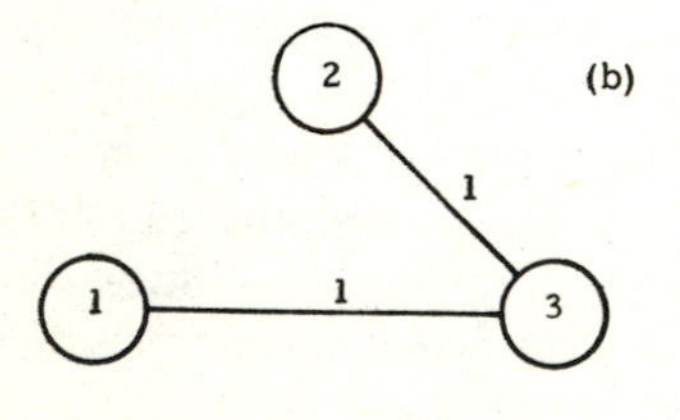

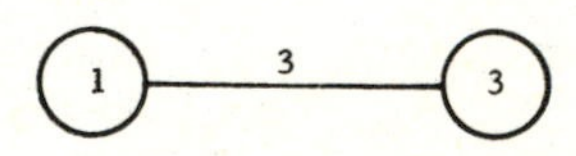

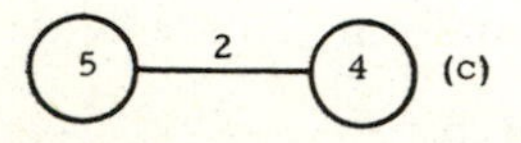

Each of the above is synthesized by a cycle,

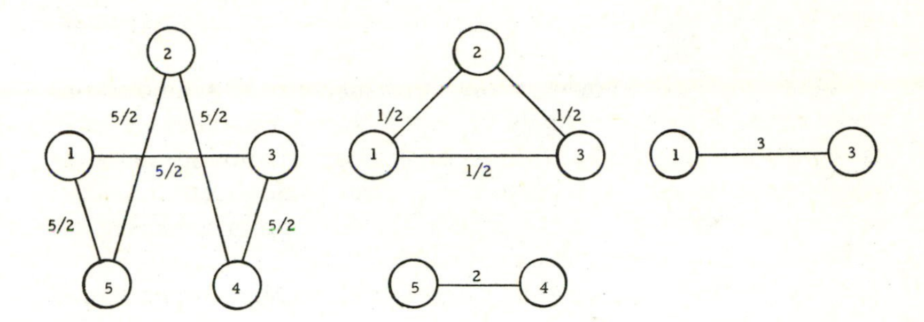

and these are superposed to give a minimal satisfactory network

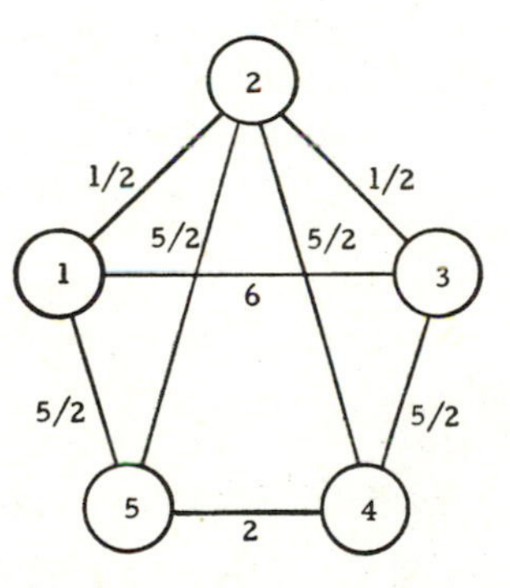

Note that in synthesizing a uniform requirement tree we may use any cycle passing through the nodes in any order. For example, (a) could have been synthesized by a cycle with capacity 5/2.

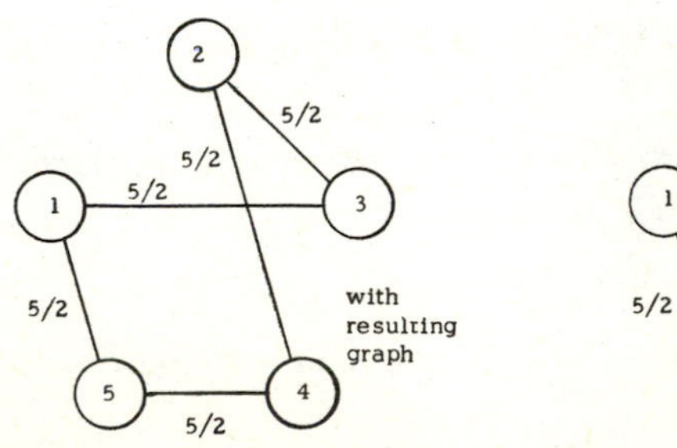

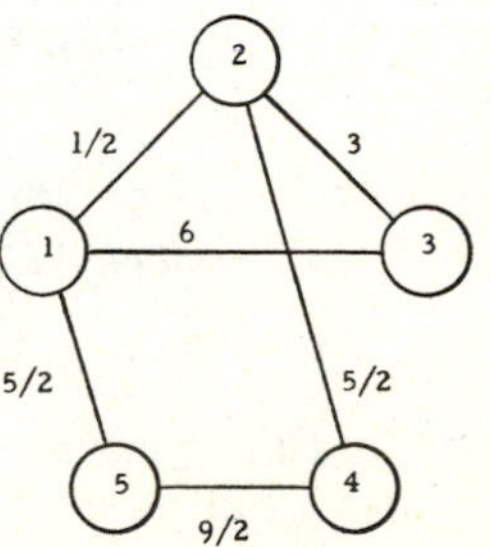

Similarly, any convex combination of cycles can be used, i.e., if β_1, β_2, β_3, . . . , are the capacities of a link in various cycles, then the graph with $\lambda_1\beta_1 + \lambda_2\beta_2, \ldots, \Sigma\lambda_i = 1$ is also a minimal synthesis and can be used.

What we have shown is that the method of synthesis will produce $f_{ij} \geqslant r_{ij}$. On actually checking the two networks synthesized above, we find that the second meets all requirements in the dominant tree exactly, while the first gives some excess flows–but of course at no cost in capacity. We shall take up first the problem of getting as much excess as possible and then the problem of exactly meeting requirements in the dominant requirement tree.

Note that the numbers u_i defined above are the ones that determine C_L, not the requirements. Consequently, once the u_i are determined, all r_{ij} can be revised upward to $\bar{r}_{ij} = \min(u_i, u_j)$ without affecting C_L (and clearly no further increase is possible on any arc without affecting C_L). If the new requirements $\bar{r}_{ij}$ are now synthesized they will be met exactly, i.e., the resulting flows $\bar{f}_{ij}$ satisfy $\bar{f}_{ij} = \bar{r}_{ij}$, for $\bar{f}_{ij} > \bar{r}_{ij}$ would necessitate, at either N_i or N_j, a larger u and hence a larger C_L. Also, the synthesized network has for the same reason the following property: let f'_{ij} be the flows provided by any other minimal capacity network satisfactory with respect to the original r_{ij}, then

$$\bar{f}_{ij} \geqslant f'_{ij} \qquad \text{(all } i, j.\text{)},$$

i.e., the network obtained by revising the requirements to $\bar{r}_{ij}$ and then synthesizing provides, at no cost in total capacity, more (or the same) flow between *every* pair of points as does any other satisfactory minimal network. More flow between *any* pair of points can be bought only by increasing total capacity.

We can summarize this property of uniform dominance in the following theorem which involves $u_i = \max_j r_{ij}$ and $\overline{C}_L = \frac{1}{2}\Sigma_{i \neq j} b_{ij}$:

THEOREM 3: *Given requirements r_{ij}, there is a satisfactory network $\overline{N}$ having capacity $\overline{C}_L$ and giving flows*

$$\bar{f}_{ij} = \min(u_i, u_j) \qquad (\text{all } i \neq j),$$

while if f_{ij} are the flows from any other satisfactory network N, then *either N's total capacity C satisfies*

$$C > \overline{C}_L \quad \text{or} \quad f_{ij} \leqslant \bar{f}_{ij} \qquad (\text{all } i \neq j).$$

We now turn to the problem of exactly meeting the requirements in the general case and, of course, at minimal capacity. We already know how to secure flows $f_{ij} \geqslant$ the requirements r_{ij}. To secure the opposite inequality, it is only necessary, after decomposing the original requirements into a sum of uniform requirement trees, to synthesize each uniform tree so that the links of the tree represent not only requirements, but also minimal cuts of the synthesized network. For instance, in our example the cut tree of the cycle used in synthesizing the uniform requirement of 5 (Figure 9) does not have a cut of capacity 5 separating 1, 2 and 3 from 4 and 5, as it would if it were a cut tree. However the synthesis by the cycle 1, 3, 2, 4, 5, 1 does have the required cut

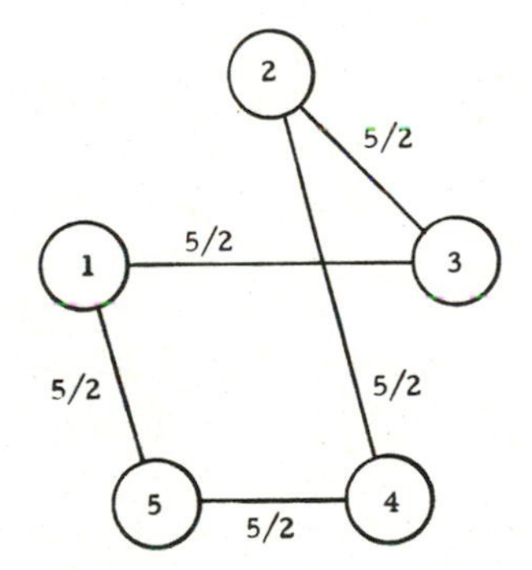

tree. In superposing requirement trees that are also cut trees, minimal cuts in the synthesized networks are superposed on minimal cuts to form minimal cuts. (Clearly, if $(A, \overline{A})$ is the minimal cut between N_i and N_j in one network and also in a second, then it is also a minimal cut in the superposed network.) Thus the original requirement tree is synthesized with a cut corresponding to each link. Hence the flow f_{ij} between two N_i and N_j satisfies

$$f_{ij} \leqslant \min(r_{ik}, r_{kl}, \ldots, r_{pj}),$$

(where the r_{ik}, etc., are in T), for each represents a cut separating N_i from N_j, i.e., if one can synthesize a uniform tree with desired cut tree, then the original requirements tree can be synthesized as a cut set tree, which results in exact synthesis.

To synthesize a given uniform tree as a cut tree is, however, quite easy. We give the following rule:

If T' is a uniform requirement tree, it is synthesized as follows:

Preliminaries: Label all arcs in T' with a zero, and choose any node as starting point. Label this node with 1.

In what follows, when we speak of labeling a node we mean to assign the first node labeled the number 1, the second node labeled the number 2, etc.

(1) Find the arc with smallest label v incident to the starting node (any one will do if there are several).

(1a) If $v \neq 0$, you must label the node.

(1b) If $v = 0$, you may or may not label the node.

(2) Proceed over the arc mentioned in (1) to the next node, increasing the arc label by 1.

(3) Continue this process until you return to your starting node and find all incident arcs labeled 2. (This will happen.) Then stop.

We assert that at this point you will have traversed all arcs of T' exactly twice (all will be labeled 2), all nodes will be labeled, and that the cycle consisting of arcs of capacity $\beta/2$ with the nodes taken in order of their labeling (and then returning to the starting node) is a synthesis of the desired cut tree.

Applying this process to the uniform 5 tree in our example gives (among others) the cycle used in Figure 9 which resulted in exact synthesis.

Proving that this general procedure works is rather tedious, so we give the following procedure which is a specialization of the one above and whose properties are more easily verified.

Suppose T' is a uniform requirement tree whose edges all have value β. We may construct a cycle λ, through all the nodes of T', such that λ has T' as a cut tree, as follows. Label any node of T' with the number 1. Then repeat the following step until T' is completely labeled: If the last label was m, then label with $m + 1$ any unlabeled node which is then adjacent to node m, if such an unlabeled node exists; if none exists, label with $m + 1$ an unlabeled node adjacent to a labeled node with the largest label possible. When T' is completely labeled, we define λ to go from node 1 to node 2 to node 3, etc., and finally to return to node 1. We let every edge of λ have capacity $\beta/2$.

Proof: To see that T' is indeed a cut tree of λ, consider any edge ϵ of T'. Suppose the two nodes of ϵ are labeled i and j, with $i < j$. Let k be the largest label which occurs in the component of $T' - \epsilon$ which contains j. Then the minimal cut whose edges are $\overline{j-1, j}$ and $\overline{k, k+1}$ corresponds to ϵ. (If $j = 1$, then $j - 1$ means the largest label, and if $k =$ the largest label, then $k + 1$ means 1.)

We have been discussing the problems of realizability, analysis and synthesis of multi-terminal network flows, where a pair of nodes is selected as the source and the sink and all other nodes serve as intermediate nodes. At a given time, only one pair of nodes is selected to act as the source and the sink. These problems are not to be confused with multi-commodity flow problems where many pairs of sources and sinks are present, and each source has its special kind of flow to its sink. All kinds of flows are simultaneously present in the network, and they share the same arc capacity. For such problems, the reader is referred to Hu [8].

APPENDIX

The following algorithm is to construct a maximum spanning tree and to check whether a given symmetrical matrix is realizable during the construction of the tree. If one wants only the maximum spanning tree, then the algorithm in [9] or [11] is more efficient. Geometrically, the algorithm is to choose the longest arc (the largest number), and then the next longest arc and so forth (arcs can be chosen only if they form a subtree). This will result in many disconnected subtrees. For each of the subtrees, we check that condition (3) is satisfied. Because the length of the arcs is monotonically decreasing, any arc δ which connects nodes belonging to a single subtree but did not do so until the arc ν was selected should have the same length as v. At the end, the maximum spanning tree is formed when all the subtrees are connected.

All this boils down to the following simple arithmetical steps in the requirement matrix.

The requirement matrix has a border row on the top and a border row in the leftmost column to indicate the ith row or jth column, as shown in Table A 1.

In what follows, r_{ij} means a matrix element in a row with border element i and column border element j. The algorithm is then the repetition of the following two steps:

Step 1. Select the largest number in the matrix proper that has not been selected or crossed out. (In the beginning, no number has been selected or crossed out.) Let this number be r_{ij}. Make a check mark in its box. If $p = \min(i, j)$ and $q = \max(i, j)$, change all q's in both borders to p's. For example, if r_{35} is chosen, then change ⑤ in the border row and column into ③.

Step 2. Consider all entries (not yet crossed out or selected) whose border entries are both p. If they are equal to the last entry selected, cross them out and return to step 1. If even one of them is not equal to the last entry selected, the matrix is not realizable.

Step 1 and Step 2 are repeated until $n - 1$ numbers are chosen. If this can be done, the matrix is realizable.

Take the following table, for example.

		①	②	③	④	~~⑤~~ ③	⑥
	①	d	7	5	3	6	4
	②		d	6	7	5	3
	③			d	3	9✓	4
	④				d	8	4
③	~~⑤~~					d	5
	⑥						d

TABLE A1

Step 1. Select $r_{35} = 9$ and change ⑤ into ③.

Step 2. The only number r_{33} is 9 itself, so no crossing out is required.

Step 1. Select $r_{43} = 8$ and change ④ into ③ . The result is shown in the following table:

	①	②	③	③	③	⑥
①	d	7	5	3	6	4
②		d	6	7	5	3
③			d	~~3~~	9✓	4
③				d	8✓	4
③					d	5
⑥						d

TABLE A2

Step 2. Check if 3 = 8? As $3 \neq 8$, we know that the matrix is not realizable, but to illustrate the algorithm, we shall continue checking. Cross out 3.

Step 1. Select $r_{12} = 7$ ($r_{23} = 7$ can equally well be chosen). Change ② to ① .

Step 2. Check $r_{11} = 7$?

Step 1. Select $r_{23} = 7$ and change all ③ 's into ① as shown below.

	①	①	①	①	①	⑥
①	d	7✓	5	3	6	4
①		d	6	7✓	5	3
①			d	~~3~~	9✓	4
①				d	8✓	4
①					d	5
⑥						d

TABLE A3

Step 2. Check if 5 = 7? 6 = 7? 3 = 7?, and cross out 5, 6, 3, etc., as shown in Table A 4.

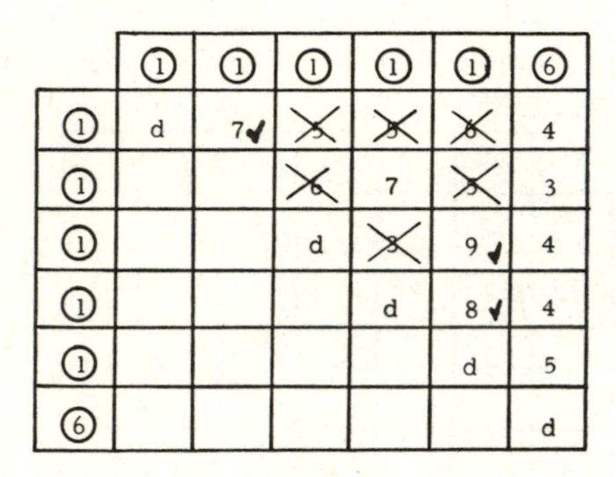

	①	①	①	①	①	⑥
①	d	7✓	~~5~~	~~3~~	~~6~~	4
①			~~6~~	7	~~5~~	3
①			d	~~3~~	9✓	4
①				d	8✓	4
①					d	5
⑥						d

TABLE A4

Step 1. Select 5 in the sixth column and change ⑥ into ①.

Step 2. Check if all elements in the sixth column equal 5.

It may be noted that if one rearranges rows and columns such that rows and columns with same labeling are next to each other, then the matrix is in Mayeda's partitioned form.

REFERENCES

1. Berge, C., *The Theory of Graphs*, translated by Alison Doig. New York: Wiley, 1962.
2. Chien, R. T., "Synthesis of a Communication Net," *I.B.M. J. Res. Develop.*, **3** (1960), 311–320.
3. Ford, L. R., and D. R. Fulkerson, "Maximal flow through a network," *Canad. J. Math.*, **8** (3) (1956), 399–404.
4. ——, "A simple algorithm for finding maximal network flows and an application to the Hitchcock Problem," *Canad. J. Math.*, **9** (2) (1957), 210–218.
5. ——, *Flows in Networks*, Princeton: Princeton University Press, 1962.

6. Gomory, R. E., and T. C. Hu, "Multi-terminal network flows," *J. of SIAM*, **9** (4) (1961), 551–570.

7. ——, "An Application of Generalized Linear Programming to Network Flows," J. of SIAM, **10** (2) (1962), 260–283.

8. Hu, T. C., *Integer Programming and Network Flows*, Reading, Mass.: Addison-Wesley, 1969.

9. Kruskal, J. B., Jr., "On the Shortest Spanning Subtree of a Graph and the Traveling Salesman Problem," *Proc. Amer. Math. Soc.*, **7** (1956), 48–50.

10. Mayeda, W., "Terminal and Branch Capacity Matrices of a Communication Net," IRE. *Transactions on Circuit Theory*, **CT-7** (1960), 261–269.

11. Prim, R. C., "Shortest Connection Networks and Some Generalizations," *Bell System Tech. J.*, **36** (1957), 1389–1401.

AUTHOR INDEX

This index covers MAA Studies Volume 11 (Studies in Graph Theory, Part I, pages 1 to 199) and MAA Studies Volume 12 (Studies in Graph Theory, Part II, pages 201–413).

SUBJECT INDEX

This index covers MAA Studies Volume 11 (Studies in Graph Theory, Part I, pages 1 to 199) and MAA Studies Volume 12 (Studies in Graph Theory, Part II, pages 201–413).